# Well Test Analysis

## Reservoir Engineering

James Daniel

# CONTENTS

## Contents

3

# Introduction to Reservoir Engineering

The primary focus of reservoir engineering, which is a subfield of petroleum engineering, is the investigation and evaluation of oil and gas reservoirs located beneath the earth's surface. It includes a wide variety of techniques, fundamentals, and methodologies that are aimed at describing, modeling, and predicting the behavior of reservoirs in order to maximize the recovery of hydrocarbon resources.

The primary goal of reservoir engineering is to optimize the recovery of oil and gas from reservoirs while also ensuring that the project will be economically viable and reducing the environmental impact as much as possible. The goal of the work done by reservoir engineers is to gain an understanding of the physical characteristics, fluid flow mechanisms, and dynamic behavior of reservoirs, which enables them to make educated decisions regarding the management of reservoirs and the production strategies to be implemented.

Key areas of focus in reservoir engineering include:

1. Reservoir Characterization: In this step, you will collect and analyze data regarding the geological, petrophysical, and hydrodynamic features of the reservoir. Methods such as well logging, seismic imaging, and core analysis are utilized in order to ascertain the spatial extent of the reservoir, as well as the properties of the reservoir rock, the fluid saturations, and the connectivity.

2. Reservoir Simulation: Throughout this stage of the process, you will be tasked with gathering and analyzing data pertaining to the geological, petrophysical, and hydrodynamic characteristics of the reservoir. In order to determine the spatial extent of the reservoir, techniques such as well logging, seismic imaging, and core analysis are applied. Also, the characteristics of the reservoir rock, the fluid saturations, and the connectivities are also determined with the help of these techniques.

3. Enhanced Oil Recovery (EOR): Engineers that work in reservoirs search for other ways to boost oil recovery in addition to the traditional

approaches. EOR techniques, including water flooding, gas injection (such as carbon dioxide or nitrogen), thermal techniques (such as steam injection), and chemical flooding, are designed to modify the fluid behavior and displacement mechanisms of the reservoir in order to maximize ultimate recovery.

4. Well Testing and Production Analysis: Engineers that work in reservoirs devise and conduct well tests in order to evaluate the features and functionality of reservoirs. They estimate reservoir parameters such as permeability, skin factor, and reservoir pressure, which are essential for the management of both wells and reservoirs based on their interpretation of pressure and flow rate data.

5. Reservoir Management: Engineers who specialize in reservoir development work closely with geoscientists, production engineers, and economists to devise the most effective strategies for reservoir production, field management, and field development. In order to maximize the recovery of hydrocarbons, it is necessary to determine the well sites, production rates, injection schemes, and monitoring procedures.

Petroleum engineers work to maximize the extraction of oil and gas from subsurface reservoirs by applying the principles and practices of reservoir engineering. In doing so, they take into account the financial, environmental, and technical concerns that are associated with reservoir management.

# Purpose of Well Test Analysis

In the industry of reservoir engineering, one of the most important methods for gathering important information about the characteristics of a reservoir is known as well test analysis. It sheds light on critical metrics like permeability, porosity, and fluid flow characteristics, all of which are essential for comprehending reservoir performance and maximizing its potential.

The major purpose of well test analysis is to analyze the behavior of fluids within a reservoir by carrying out controlled experiments on the well. These tests are carried out to determine the behavior of the fluids. These experiments entail controlling the flow of fluids either into or out of the well and monitoring the pressure responses that occur as a result of these changes. Reservoir engineers are able to obtain useful information about the parameters of the reservoir by studying the pressure measurements taken over a period of time.

The permeability is an important metric that can be determined through the study of well tests. The degree to which a reservoir rock is permeable, as measured by its capacity to allow fluids to pass through it, is one of the most important factors that determines the rate at which hydrocarbons flow. The engineers are able to estimate the permeability of the reservoir by examining the pressure data collected during a well test. Knowing the permeability of the reservoir is essential for accurately estimating production rates and developing the most effective production methods.

Porosity is another essential quality, and it relates to the volume of void space that exists inside the reservoir rock and has the capacity to hold fluids. The study of well tests can shed light on the effective porosity of the reservoir, which is necessary for accurately predicting the total amount of hydrocarbons that the reservoir is capable of holding as well as the flow capacity of the rock.

The study of well tests also includes an evaluation of the flow properties of the fluid. This involves determining the kind of fluid that is present in the reservoir, such as oil, gas, or water; the fluid's composition; and how the fluid reacts when subjected to varying circumstances in the reservoir. Engineers are able to comprehend the fluid flow patterns within a reservoir by analyzing pressure responses, which might reveal the presence of borders, barriers, or compartmentalization within the reservoir.

In order to interpret the pressure data that was obtained during the well test, the analysis of the well test requires the use of a variety of mathematical and analytical approaches. These methods include pressure transient analysis, which concentrates on examining the variations in pressure over time, and derivative analysis, which assists in identifying particular reservoir characteristics. Both of these analyses can be carried out simultaneously.

The insights gathered from the analysis of well tests are essential for making educated decisions regarding the management of reservoirs and the optimization of production. They are helpful in determining the economic viability of a reservoir, planning efficient well completion and stimulation procedures, calculating reservoir reserves, evaluating well performance, and analyzing overall reservoir effectiveness.

Well test analysis is a strong tool in reservoir engineering that allows engineers to acquire data about the permeability, porosity, and fluid flow characteristics of a reservoir. In general, well test analysis is one of the most important aspects of reservoir engineering. The optimization of reservoir output and the maximization of hydrocarbon recovery both require this knowledge to a crucial degree.

The term "well test analysis" refers to one of the most important procedures that are utilized in the field of reservoir engineering. This method is utilized for the purpose of acquiring essential knowledge regarding the qualities of a reservoir. It gives insight on crucial parameters such as permeability, porosity, and fluid flow characteristics, all of which are essential for understanding reservoir performance and making the most of the potential it possesses.

The primary objective of carrying out controlled experiments on the well for the purpose of conducting well test analysis is to investigate the ways in which fluids behave within a reservoir. These examinations are carried out so that the behavior of the fluids can be determined. In the course of these experiments, you will be in charge of regulating the flow of fluids either into or out of the well, and you will be tasked with keeping an eye on how the pressure reacts to the various shifts in flow. By analyzing the pressure readings that have been collected over the course of a certain amount of time, reservoir engineers are able to acquire knowledge that is applicable to the parameters of the reservoir.

The permeability is a crucial parameter that may be figured out by analyzing the results of well tests. The degree to which a reservoir rock is permeable, which is measured by its capacity to allow fluids to pass through it, is one of the most important factors that determines the rate at which hydrocarbons flow. This capacity is measured in terms of the rock's ability to allow fluids to pass through it. When the engineers examine the pressure data that was acquired during a well test, they are able to make an educated guess as to the permeability of the reservoir. It is vital to have a thorough understanding of the permeability of the reservoir in order to establish the most efficient production methods and conduct reliable estimates of production rates.

Porosity is another vital attribute, and it refers to the volume of empty space that is present inside the reservoir rock and has the ability to hold fluids. Porosity is essential because it determines whether or not the rock can store fluids. The analysis of well tests can shed light on the effective porosity of the reservoir, which is necessary for accurately predicting the total amount of hydrocarbons that the reservoir is capable of holding as well as the flow capacity of the rock. This can be accomplished by studying the effective porosity of the reservoir.

The investigation of well tests also incorporates an analysis of the flow characteristics of the fluid being studied. This involves determining the type of fluid that is present in the reservoir, such as oil, gas, or water; the composition of the fluid; and how the fluid reacts when subjected to different conditions in the reservoir. For example, oil may react differently to different temperatures than gas may react differently than water. By evaluating pressure responses, engineers are able to gain an understanding of the fluid flow patterns that occur within a reservoir. These pressure responses may show the existence of borders, obstacles, or compartmentalization within the reservoir.

In order to interpret the pressure data that was collected during the well test, the analysis of the well test involves the application of a number of mathematical and analytical approaches. These approaches are necessary in order to interpret the data that was obtained. These methods include pressure transient analysis, which focuses on studying the variations in pressure over time, and derivative analysis, which helps in detecting unique reservoir characteristics. Both of these analyses concentrate on examining the pressure data over a period of time. These two examinations can be performed alongside one another without any problems.

For the purpose of making informed decisions on the management of reservoirs and the optimization of production, the insights that are gleaned from the analysis of well tests are absolutely necessary. They are helpful in determining the economic viability of a reservoir, planning efficient well completion and stimulation procedures, calculating reservoir reserves, evaluating well performance, and analyzing the overall effectiveness of a reservoir. In addition, they are helpful in planning efficient well completion and stimulation procedures.

The analysis of well tests is a powerful technique in the field of reservoir engineering. It enables engineers to collect data about the permeability, porosity, and fluid flow properties of a reservoir. In the field of reservoir engineering, general well test analysis is considered to be one of the most significant parts. Both the maximizing of hydrocarbon recovery and the optimizing of reservoir output require this knowledge to an essential degree in order to be successful.

Well tests are valuable tools for estimating various key reservoir properties. Let's discuss some of the properties that can be estimated from well tests:

Reservoir Pressure: Well testing gives critical information concerning reservoir pressure. The well is sealed off during pressure accumulation tests, which are also known as shut-in tests, and the pressure within is monitored as it fluctuates over time. Engineers are able to determine the average pressure in the reservoir by examining the data on pressure accumulation, which is a critical step in both the characterisation of the reservoir and the production forecasting process.

Permeability: The ability of a rock to permit the flow of a fluid is referred to as its permeability. The permeability of the reservoir can be better understood with the help of certain well tests, such as pressure transient analysis. Engineers are able to conduct an analysis of pressure transient behavior and determine the permeability of a reservoir by observing the pressure responses during production or injection.

Skin Factor: The skin factor is an indicator of the damage or enhancement that occurs around the wellbore and has an effect on the flow of fluid into the well. Tests conducted on the well, such as pressure drawdown tests, can assist in the estimation of the skin factor. Engineers are able to identify the impact of wellbore conditions on fluid flow and calculate the skin factor by conducting an analysis of the pressure decline data that is collected during production.

Reservoir Boundaries: The boundaries of a reservoir can be determined with great accuracy through the use of well testing. Monitoring the pressure responses in numerous wells is required for the various pressure transient analysis procedures, such as the pressure interference tests and the pulse tests. Engineers are able to determine the position and extent of reservoir borders through the analysis of pressure data as well as the identification of pressure communication between wells.

Fluid Properties: In addition, well testing can shed light on the characteristics of reservoir fluids, such as the oil, gas, or water that are present in the reservoir. PVT analysis, which stands for pressure-volume-temperature analysis, is often used to evaluate the properties of fluids. During well tests, engineers can determine factors such as fluid density, viscosity, compressibility, and phase behavior by measuring changes in pressure, volume, and temperature.

It's important to note that estimating these properties from well tests requires sophisticated analysis techniques and models. Various mathematical and numerical methods, such as pressure transient analysis,

10

well-test interpretation, and reservoir simulation, are used to extract valuable information from well test data. These analyses help engineers make informed decisions regarding reservoir development, production strategies, and overall reservoir management.

The process of interpreting the results of a well test is a crucial part of reservoir engineering. The goal of this process is to estimate the properties of the reservoir by doing an analysis of the pressure response that is recorded during a well test. In order to achieve a higher degree of precision in the estimated reservoir attributes, this method necessitates an iterative process in which several models and interpretations are judged in light of the data that has been gathered. The following is an overview of the iterative process that is involved in the interpretation of well tests:

1.  Data Acquisition: The process begins with conducting a well test, such as a pressure buildup or drawdown test, to obtain pressure and flow rate measurements over a specific period.

2.  Model Selection: Different models are available for well test interpretation, each based on different assumptions and mathematical formulations. The appropriate model is selected based on the reservoir characteristics and the objectives of the interpretation.

3.  Initial Interpretation: An initial interpretation is performed by matching the observed pressure and flow rate data with the model's predictions using estimated reservoir properties. This initial interpretation serves as a starting point for subsequent iterations.

4.  Sensitivity Analysis: Sensitivity analysis is conducted to identify the most influential reservoir parameters that significantly affect the match between the model and observed data. These parameters can include reservoir permeability, porosity, skin factor, and reservoir boundaries.

5.  Model Calibration: The initial interpretation is adjusted by modifying the estimated reservoir properties, such as permeability

or skin factor, based on the sensitivity analysis. This calibration process aims to improve the match between the model predictions and the observed data.

6.  Error Analysis: The interpretation process involves quantifying the error between the model predictions and the observed data. Statistical techniques, such as least-squares regression or maximum likelihood estimation, may be employed to assess the goodness-of-fit and identify any systematic discrepancies.

7.  Model Improvement: Based on the error analysis and calibration results, modifications are made to the model formulation or additional parameters are considered to refine the interpretation. This may involve incorporating more complex reservoir behavior, considering fluid flow mechanisms, or accounting for wellbore effects.

8.  Iteration and Convergence: Steps 4-7 are repeated iteratively until a satisfactory match between the model predictions and the observed data is achieved. The interpretation process converges when the estimated reservoir properties reach a stable and consistent solution.

9.  Uncertainty Analysis: Once a converged interpretation is obtained, uncertainty analysis is performed to quantify the confidence and range of uncertainty associated with the estimated reservoir properties. This analysis helps assess the reliability and robustness of the interpretation results.

10. Reporting and Decision-Making: The final interpretation results, including the estimated reservoir properties and associated uncertainties, are reported and used to make informed decisions regarding reservoir management, production optimization, and future development strategies.

Overall, the iterative process in well test interpretation consists of adjusting and refining the initial interpretation through model calibration, error analysis, and sensitivity analysis until a consistent and accurate estimation of the reservoir properties is achieved. This process continues until a consistent and accurate estimate of the reservoir properties is achieved. Because of this approach, reservoir engineers are able to get vital insights into the behavior of the reservoir and make decisions based on accurate information in order to maximize hydrocarbon output.

# Reservoir Characterization

A vital component of reservoir engineering, reservoir characterization necessitates an in-depth familiarity with the geology, petrophysical, and hydrodynamic features of a reservoir. This knowledge is necessary in order to design and optimize a reservoir. It gives vital information regarding the properties, behavior, and potential for hydrocarbon production of the reservoir. In the following, I will provide a more in-depth explanation of the methods that are utilized in reservoir characterization:

1. Well Logging: Well logging is a technique used to gather information about the subsurface formations encountered by drilling wells. Various logging tools are lowered into the wellbore to measure physical properties such as electrical resistivity, porosity, permeability, lithology, and fluid saturations. These measurements help in identifying hydrocarbon-bearing zones, determining rock and fluid properties, and evaluating the reservoir's productivity.

1. Resistivity Logging: Resistivity logging measures the electrical resistivity of the subsurface formation. It is a fundamental well logging tool used to identify the presence of hydrocarbons, differentiate between rock types, and assess reservoir quality.

- Induction Logs: The resistivity of a formation can be measured with the help of induction logs by applying the principles of electromagnetic induction. The conductivity of the formation fluids and the resistivity of the rocks in the surrounding area both have an effect on the readings. In general, the resistivity values of hydrocarbon-bearing formations are significantly higher than those of water-bearing formations.

- Laterologs: The formation's resistivity can be measured at a variety of examination depths using laterolog instruments. In order to determine the resistivity of the area immediately surrounding the wellbore, they make use of current and voltage electrodes that are kept apart by insulating materials. These logs offer a more comprehensive analysis of the resistivity of the formation at varying radial distances from the wellbore.

- Microresistivity Logs: Tools that measure microresistivity give measurements of resistivity with a high resolution. In order to collect specific information regarding thin beds, fractures, and formation heterogeneities, they use small electrodes and focussed currents.

- Imaging Logs: A spatial image of the resistivity surrounding the wellbore can be obtained with the use of resistivity imaging logs, such as azimuthal resistivity tools. They provide information regarding the direction and distribution of resistive and conductive zones, which helps in the identification of reservoir compartments and fault zones.

2. Porosity Logging: Porosity logging estimates the pore volume or the percentage of void spaces within the rock formation. It provides crucial insights into the storage capacity for hydrocarbons and the reservoir's potential.

- Density Logs: Density logging measures the bulk density of the formation by evaluating the attenuation of gamma rays emitted from a radioactive source. The density is influenced by the composition and porosity of the rock. Porous formations with higher hydrocarbon content typically exhibit lower bulk density.

- Neutron Logs: Neutron logging measures the hydrogen index or hydrogen concentration in the formation. Neutrons emitted by a radioactive source interact with hydrogen atoms present in the formation. Hydrogen is abundant in fluids like water and hydrocarbons. The measurement helps estimate the porosity and differentiates between hydrocarbon and water-bearing zones.

- Sonic Logs: The time it takes compressional (P-wave) or shear (S-wave) sound waves to pass through a formation is one of the variables that may be measured via sonic logging. The density and elastic qualities of the rocks both have an effect on the speed at which sound waves travel through the rocks. The presence of fluids in porous formations typically results in slower velocities, which makes it possible to estimate the formation's level of porosity.

3. Sonic Logging: The speed at which sound waves move through a formation is measured using a technique known as sonic logging. It is useful in determining the density of rocks, as well as their mechanical qualities and whether or not fractures are present.

- Compressional Sonic Logs: The amount of time that compressional (P-wave) sound waves take to travel is what compressional sonic logs measure. They offer data regarding the compressional velocity, which is connected to the elastic characteristics of the formation and the rock density. These logs assist in determining the lithology of the reservoir, as well as estimating its porosity and evaluating its mechanical qualities.

- Shear Sonic Logs: The amount of time it takes for shear (S-wave) sound waves to move is what shear sonic logs measure. They contribute information regarding the shear velocity of the formation, which is helpful for the assessment of rock characteristics, the detection of fractures, and the research of geomechanical processes.

- Full Waveform Sonic Logs: Complete waveform The full waveforms of both compressional and shear waves can be recorded via sonic logging. It makes it possible for sophisticated methods of analysis and interpretation to be used in order to obtain specific information about the formation.

4. Nuclear Logging: When conducting nuclear logging, radioactive sources and detectors are utilized in order to quantify various features of the formation, such as the porosity and lithology of the rock.

- Gamma Ray Logs: The amount of natural gamma radiation that is emitted by the formation is what gamma ray logs measure. The composition of the formation, as well as its lithology and clay concentration, all have an impact on the measurement. Gamma ray logs are useful for determining the lithology of the rock, finding shale intervals, and assessing porosity.

- Spectral Gamma Ray Logs: The intensity of gamma rays at various energy levels can be measured using pectral gamma ray logs. This makes it possible to determine the lithology of the formation and identify variations in its mineralogical composition, as well as to identify and quantify certain radioactive elements that are already present in the formation.

- Neutron Logs: As was mentioned previously, neutron logs are useful for estimating porosity because they measure the amount of hydrogen that is present in the formation. They do this by utilizing a radioactive source that emits fast neutrons and detectors to collect the thermalized neutrons that are produced as a result of the neutrons interacting with hydrogen atoms.

5. Density Logging: The purpose of determining rock characteristics and fluid saturations by density logging is to determine the bulk density of the formation.

- Density Logs: The attenuation of gamma rays is measured with the help of density logs, which contain a radioactive source and detectors. The measurement offers insight into the formation's bulk density, which is determined by the composition of the rock as well as the porosity of the rock. It assists in distinguishing between lithologies, providing an estimate of porosity, and determining the fluid saturations.

   - Compensated Density Logs: To improve the precision of density readings and take into account environmental factors such as borehole size and mud characteristics, compensated density logs make the necessary adjustments. Estimates of the formation's density and porosity can be more accurately derived from these logs.

Reservoir engineers are able to construct a full picture of the subsurface reservoir by using these logging techniques in combination with one another and integrating those findings with other data. The data that is gathered by well logging is essential for reservoir modeling, simulation, and effective reservoir management. This data helps to characterize the reservoir's lithology, porosity, permeability, fluid saturations, and mechanical properties.

# Seismic Imaging:

The creation of a detailed subsurface image of the reservoir can be accomplished by the use of seismic waves in seismic imaging. It contributes to a better knowledge of the structure of the reservoir, the mapping of subsurface features, and the identification of possible hydrocarbon traps. During seismic surveys, controlled shocks or explosions are used to create surface activity, and the waves that are reflected back from the surface are recorded. After that, cutting-edge processing methods are employed to generate a two-dimensional or three-dimensional image of the subsurface, which provides invaluable insights into the spatial extent and connectivity of the reservoir.

## Reflection Seismic:

Reflection seismic is a method that is frequently utilized in the process of characterizing reservoirs. This method makes use of controlled sources and receivers in order to record and analyze the reflected seismic waves that are generated by subsurface interfaces. Imaging the subsurface geology and gaining significant insights into reservoir attributes are both made possible because of its important function. I'll elaborate on reflection seismic and its primary components in the following:

1. Acquisition: In order to collect data by the method of reflection seismicity, seismic sources and receivers need to be placed in a carefully orchestrated manner. These seismic waves can be generated by vibroseis trucks, explosives, or air cannons, all of which penetrate the subsurface and produce seismic activity. The receivers, which may also be referred to as geophones or hydrophones, are installed in boreholes or in strategic locations on the surface in order to capture the reflected waves.

2. Seismic Waves: Seismic waves are a type of elastic wave that travel through the subsurface of the Earth and transmit information about the geological structures and rock qualities they pass through. In reflection seismic, the two basic forms of seismic waves that are used are as follows:

    - P-waves (Primary Waves): They are compressional waves that move through solids, liquids, and gases all at the same time. These are the first arrivals recorded by the receivers, and the information they provide regarding the subsurface velocity and lithology can be found in these arrivals.

    - S-waves (Shear Waves): These are transverse waves that travel at a slower speed and can only travel through solids. They offer further insight into the characteristics of the rock, particularly its shear strength and anisotropy.

3. Reflection and Refraction: Due to the fact that they have different acoustic qualities, they experience both reflection

and refraction. Waves are said to have refraction when they bend as they go through strata with different velocities, but waves that bounce back towards the surface are said to have reflection. Reflection happens when waves return to the surface after being deflected by an obstacle.

4. Data Processing: After the seismic data has been obtained, it is subjected to intensive processing in order to improve the data's quality and derive information that is meaningful. The following are the stages of the data processing:

   - Deconvolution: Removes the effects of the seismic wavelet from the recorded data, improving resolution.

   - Noise Removal: Eliminates unwanted noise sources, such as environmental and acquisition-related noise.

   - Migration: Corrects the positioning of the seismic events to their actual subsurface locations, resulting in a more accurate image.

   - Velocity Analysis: Estimates the subsurface velocity distribution, which is crucial for accurate imaging and interpretation.

## Seismic Attributes:

Quantitative measurements that are obtained from seismic data and provide extra information about the features of the reservoir are referred to as seismic attributes. They are put to use in the process of interpreting geological features and characterizing the subsurface. In the following, I'll provide some further information on seismic features and the significance of these attributes in reservoir characterization:

1. Types of Seismic Attributes: The seismic data can be used to derive a wide variety of seismic properties and characteristics. The following are examples of qualities that are frequently used:

   - Amplitude: Measures the strength of the seismic reflections and can indicate lithology and fluid presence.

   - Frequency: Provides information about the size and geometry of subsurface features.

   - Phase: Determines the relative timing of seismic reflections and aids in identifying fault locations.

   - Coherence: Measures the similarity between seismic traces and helps identify stratigraphic boundaries and faults.

   - AVO (Amplitude Variation with Offset): Quantifies the changes in seismic amplitudes with varying source-receiver distances and can indicate fluid presence and reservoir properties.

   - Spectral Decomposition: Analyzes the seismic data in different frequency bands to highlight specific geologic features.

2. Interpretation and Reservoir Analysis: The analysis and interpretation of seismic characteristics is an extremely important part of reservoir exploration. They are useful in:

   - Identifying and mapping reservoir structures, such as faults, fractures, and stratigraphic features.

   - Estimating reservoir properties, including porosity, fluid saturation, lithology, and permeability, by establishing correlations between seismic attributes and well log data.

   - Delineating hydrocarbon-bearing zones and potential

traps.

- Planning well locations, field development strategies, and production optimization based on the insights gained from seismic attribute analysis.

Reservoir engineers are able to gain a more in-depth understanding of the subsurface properties and make more educated decisions pertaining to reservoir management and production strategies thanks to seismic attributes, which are an essential tool for reservoir characterization. Seismic attributes are also known as seismic signatures. When seismic features are combined with those from other data sources, such as well logs and core analysis, a complete image of the reservoir can be created. This, in turn, makes it possible to facilitate hydrocarbon recovery that is both efficient and effective.

# Core Analysis:

Rock samples in the form of cylinders called cores can be obtained by drilling wells. The inspection of these samples in great detail in a laboratory is a necessary part of the core analysis process, which is used to evaluate a variety of rock properties and the behavior of fluids. Core analysis approaches include:

## Petrography:

The study of thin sections of core samples under a microscope is the primary step in the petrographic method of determining the characteristics of a reservoir. A wafer-like slice of rock is used to create a thin portion, which is then polished and put on a glass slide. Reservoir engineers are able to detect and define the various types of minerals and rock textures that are present in the reservoir by examining these thin slices using a petrographic microscope.

The first step in petrography is preparing the samples to be analyzed. A representative core sample is chosen after careful consideration and then sliced into a tiny block. After that, a specialized cutting machine is used to split off the block into extremely thin slices, with a thickness of approximately 30 micrometers on average.

Following this step, the resulting thin section is mounted on a glass slide and then ground to get the desired accurate thickness as well as a smooth, flat surface.

The petrographic microscope is used to do an examination of the thin section using both transmitted and reflected light. With a microscope, one can examine the mineralogical makeup of a rock, as well as its texture, grain size, and the arrangement of the grains within the rock. The ability to identify individual minerals relies on their unique optical features, which can include color, birefringence, and crystal structure, amongst others. Petrography is useful for determining the kind of rock, such as sandstone, limestone, shale, or carbonate, and it also offers insights on the depositional environment and digenetic processes that have altered the rock.

## Porosity and Permeability Measurement:

The amount of pore space in the rock and the connectivity of those pores, as well as the capacity of fluids to move through the reservoir, are both determined by the porosity and permeability of the rock. Porosity and permeability are two of the most important parameters in reservoir characterization.

The percentage of vacant space, also known as pores, that is contained inside a rock's volume is referred to as its porosity. Because it reflects the amount of space the reservoir has available for storing hydrocarbons, it is an extremely important statistic. There are a few different ways that porosity can be measured, some of which are:

> 1. Core Plug Analysis: Small cylindrical plugs are extracted from core samples, and their volume is measured. The plugs are then dried, and their dry weight is measured. By comparing the initial volume with the dry weight, the total porosity can be calculated.

2. Gas or Liquid Displacement: A sample of the rock is saturated with a gas (e.g., nitrogen) or liquid (e.g., mercury), and the volume of the fluid required to fill the pore space is measured. This method helps determine the total porosity as well as differentiating between effective and ineffective porosity.

On the other hand, a rock's permeability can be measured by determining how easily fluids can pass through it. It is dependent on the interconnectivity of the pores as well as their size. Many methods, including the following, can be used to determine a material's permeability:

1. Core Flooding: A core plug is placed in a specially designed apparatus where fluids are injected at controlled pressures. The rate of fluid flow through the core is measured, and permeability is calculated using Darcy's law.

2. Pulse Decay or Constant-Rate Tests: These tests involve measuring the pressure drop across a rock sample as fluid is injected. By monitoring the pressure decay over time, permeability can be determined.

## Saturation Analysis:

Saturation analysis is a method that is used to determine the fluid saturations in the rock, such as the amount of water and hydrocarbons that are present. It is essential for determining the amount of the reservoir's hydrocarbons that are potentially recoverable

Measurements taken in a laboratory and study of well logs are often both components of a saturation analysis. Fluid samples taken from the reservoir are subjected to a variety of tests in the laboratory, including centrifugation, extraction using the Dean-Stark method, and chemical analysis of the samples. These methods make it possible to determine the levels of oil saturation and water saturation, among other types of fluid saturation.

Saturation analysis relies heavily on well log interpretation due to its importance. Indirect measurements of fluid saturations can be obtained by the use of a variety of well logs, such as neutron logs and resistivity logs, which are based on the electrical characteristics and responses to radiation exhibited by the formation. The core measurements are used to calibrate these logs, which are then used to estimate the fluid saturations throughout the reservoir.

## Rock Mechanical Properties:

Rocks mechanical characteristics are vitally important to the characterization of reservoirs because of the direct influence they have on the behavior of reservoirs during production activities. The rock's strength, elasticity, and brittleness are three of its most important mechanical qualities.

1. Rock Strength: Rock strength is defined as the reservoir rock's capacity to withstand the application of external forces without undergoing irreversible deformation or breaking down. It is essential for figuring out the stability of wellbores, as well as the design of drilling and completion procedures. The strength of rocks is commonly evaluated in a laboratory using unconfined compressive strength (UCS) tests or triaxial tests, in which rock samples are subjected to increasing pressure until failure occurs. Other types of rock strength evaluations also exist.

2. Rock Elasticity: The response of the reservoir rock to an externally applied stress is characterized by the elastic properties of the rock, such as Young's modulus and Poisson's ratio. During production activities, these features have an effect on the deformation and stress distribution that occurs inside the reservoir. Tests in the laboratory, such as ultrasonic velocity measurements and acoustic wave propagation tests, are used to assess the properties of an elastic material.

3. Rock Brittleness: The tendency of the rock to fracture rather than deform when subjected to stress is what is meant by the term "brittleness." It is a key factor to consider while doing hydraulic

fracturing, sometimes known as fracking. Many methods, such as mineralogical analysis, rock mechanical tests, and well log analysis, can be utilized in order to arrive at an estimation of brittleness.

The evaluation of these rock mechanical properties enables reservoir engineers to gain insights into the behavior of the reservoir during production, well stability, and the potential for fracture propagation. These insights are then utilized to assist in the development of strategies for reservoir management and production optimization.

# Fluid Analysis:

Fluid samples obtained from wells are analyzed to determine the properties of reservoir fluids, such as crude oil, natural gas, and formation water. Fluid analysis techniques include:

1. PVT (Pressure-Volume-Temperature) Analysis:

PVT analysis is a crucial component of reservoir characterization that involves studying the behavior of reservoir fluids (oil and gas) under different pressures and temperatures. It provides critical information about the phase behavior, fluid properties, and volume changes of hydrocarbons within the reservoir. The key objectives of PVT analysis include:

- Determining Phase Behavior: PVT experiments help identify the phase behavior of reservoir fluids, including the critical properties, phase envelopes, and phase transitions. This information is essential for estimating the volume of hydrocarbons in different phases (oil, gas, and condensate) and understanding their behavior during production.

- Estimating Fluid Properties: PVT analysis enables the

estimation of various fluid properties, including oil and gas density, viscosity, compressibility, formation volume factor (FVF), and gas-oil ratio (GOR). These properties are crucial for reservoir modeling, flow simulation, and reserves estimation.

- Reservoir Simulation: PVT data is incorporated into reservoir simulation models to accurately predict fluid flow behavior, pressure changes, and production performance. Simulation helps optimize production strategies, such as well placement, production rates, and recovery methods.

PVT analysis involves collecting fluid samples from the reservoir and subjecting them to controlled laboratory experiments. The fluid samples are exposed to different pressures and temperatures, and measurements are taken to determine properties such as bubble point pressure, oil formation volume factor, oil viscosity, gas solubility, and more. These measurements provide valuable insights into the fluid behavior and properties necessary for reservoir characterization and management.

2. Gas Chromatography:

Gas chromatography is a technique used to separate and analyze the individual components of natural gas. It provides detailed information about the composition, concentration, and physical properties of the gas components. Gas chromatography is widely used in reservoir characterization for the following purposes.

- Component Identification: Gas chromatography allows the identification and quantification of various hydrocarbon components present in natural gas, including methane, ethane, propane, butane, and higher hydrocarbons. It also detects non-hydrocarbon gases like nitrogen, carbon dioxide, and hydrogen sulfide. Knowing the gas composition helps determine the gas quality, potential impurities, and its suitability for different applications.

- Gas Analysis and Properties: Gas chromatography provides data on various properties of natural gas components, such as molecular weight, specific gravity, heat content, and adiabatic flame temperature. These properties are essential for assessing gas quality, estimating reserves, and designing gas processing facilities.

- Gas Condensate Analysis: In reservoirs where gas condensates are present, gas chromatography is employed to analyze the liquid hydrocarbon components in the gas condensate. This analysis helps determine the liquid yield, composition, and physical properties of the condensate, which are crucial for reservoir characterization and production planning.

In gas chromatography, one separates the various components of a gas by employing a column that is filled with a stationary phase. When the gas mixture travels along the column, the different components begin to separate based on the degree to which they are attracted to the stationary phase. Following the separation of the components, they are then detected and studied, which results in a thorough understanding of the properties and composition of the gas.

3. Reservoir Fluid Saturation:

Reservoir fluid saturation refers to the fraction of hydrocarbons (oil and gas) and water present within the reservoir rock. Determining the fluid saturation is vital for estimating reserves, understanding the reservoir's potential, and designing effective production strategies. Several methods are employed to evaluate reservoir fluid saturation, including:

- Well Logging: Well logging tools, such as resistivity and neutron porosity logs, can differentiate between hydrocarbons and water based on their electrical properties. These logs provide information about the fluid saturations in the reservoir rock.

- Pressure Transient Analysis: Pressure transient tests, such as well tests and pressure buildup tests, can provide insights into the fluid saturations by analyzing the pressure response of the reservoir to production or shut-in periods.

- Core Analysis: Core analysis techniques, such as Dean-Stark extraction or solvent extraction, can be used to determine the oil and water saturations within the rock samples obtained from drilling wells.

- Production Analysis: Analyzing the fluid production rates and composition from producing wells can give indications of the fluid saturations and connectivity within the reservoir.

Reservoir engineers are able to make an educated guess on the fluid saturations present within the reservoir by combining data obtained from well logging, pressure transient analysis, core analysis, and production analysis. This information is essential for estimating reserves, figuring the recovery factors, and optimizing production techniques like well placement, perforation intervals, and gas injection or waterflooding strategies.

PVT analysis, gas chromatography, and reservoir fluid saturation analysis all contribute significantly to the overall process of characterizing a reservoir by yielding information about the fluid's behavior, composition, and distribution within the reservoir. The utilization of these methodologies facilitates the formation of well-informed judgments on reservoir management, production optimization, and reserve assessment.

Reservoir engineers are able to acquire a comprehensive understanding of the reservoir's spatial extent, lithology, porosity, permeability, fluid saturations, and connectivity by utilizing the aforementioned methodologies. This data is used as the foundation for modeling and simulating reservoirs, as well as for making decisions concerning reservoir management strategies, production optimization, and increased oil recovery methods.

# Pressure Transient Analysis

In the fields of reservoir engineering and well testing, a specialized method known as pressure transient analysis is utilized to evaluate the pressure data received during well tests. In order to get important insights on reservoir parameters, fluid flow behavior, and the performance of the well and reservoir, it requires evaluating the changes in pressure over time.

The flow rate of fluids (such as oil, gas, or water) is adjusted during a well test, and the accompanying pressure response is measured either at the wellbore or at certain sites within the reservoir. Analyzing the behavior of pressure as it changes over time is the primary emphasis of pressure transient analysis. This analysis takes into account a variety of elements, including fluid flow dynamics, reservoir geometry, and rock characteristics.

Estimating the essential reservoir parameters that influence fluid flow is the fundamental purpose of pressure transient analysis. These properties include permeability, porosity, reservoir limits, and the presence of any near-wellbore damage or stimulation effects. In addition to this, it assists in comprehending the interconnectedness and compartmentalization of reservoirs.

In order to gain useful information from pressure data, conducting a pressure transient analysis necessitates applying several mathematical models and methods of analysis to the information in question. The radial flow model is the one that is utilized the vast majority of the time. This model presupposes that fluid flow happens radially from the wellbore into the reservoir. This model operates on the presumption that the reservoir in question is both homogenous and isotropic.

Reservoir engineers are able to determine the permeability of the reservoir by monitoring pressure transients. Permeability is a measurement of the rock's ability to transport fluids from one location to another. Engineers are able to evaluate the productivity potential of the reservoir and build appropriate production methods thanks to the pressure response, which offers significant information on the size and connectivity of the reservoir drainage area.

In addition, pressure transient analysis is helpful in analyzing the presence of near-wellbore effects, such as wellbore storage, skin damage, or formation damage brought on by drilling or completion activities. This can be done by comparing the pressure history of the well to a reference well. These impacts can have a considerable impact on the performance of the well, and they need to be correctly assessed in order to maximize production and decide whether or not corrective actions are necessary.

In addition, pressure transient analysis is helpful in describing the borders of the reservoir and locating any potential compartmentalization that may exist inside the reservoir. This information is extremely important for the management of reservoirs since it plays a role in decision-making about well spacing, reservoir development, and enhanced oil recovery methods.

In order to properly interpret the results of a pressure transient study, it is necessary to compare the observed pressure data with mathematical models using specialized software and iterative methods. The engineers are able to assess the parameters of the reservoir, analyze the performance of both the well and the reservoir, and then make educated decisions regarding the management plans for the reservoir.

In a nutshell, reservoir engineering's pressure transient analysis is a strong tool that involves interpreting pressure data obtained during well testing. This data can be obtained from the well. It gives extremely helpful insights into reservoir characteristics, fluid flow behavior, near-wellbore effects, and performance of the reservoir. Engineers are able to enhance the amount of hydrocarbons recovered by optimizing reservoir development through the proper analysis of pressure transients.

# The Radial Flow Model

In the field of pressure transient analysis, a popular type of mathematical model is known as the radial flow model. It is predicated on the idea that fluid movement inside of a reservoir happens radially outward from the wellbore into the rock formation that is around it. This model assumes that the reservoir under consideration is homogeneous and isotropic, which means that the reservoir possesses the same characteristics in any direction you look at it.

The reservoir is depicted in the radial flow model as having the shape of a cylinder and is situated all around the wellbore. It takes into account the flow of fluids in radial directions and makes the assumption that the reservoir has an extent that is unlimited and has no limitations or impediments that could impede the flow of fluids.

In order to put the radial flow model into practice, the pressure data produced from a well test must first be examined using a number of different methods and pieces of software. The following is an outline of the primary stages involved in predicting the behavior of radial flow and extracting information that is meaningful:

1. Build a pressure history: The pressure data collected during the well test is organized to create a pressure history. This history typically includes measurements of pressure over time at different locations within the reservoir.

2. Identify the radial flow period: The pressure data is examined to identify the period during which radial flow dominates. This period occurs after the well has been shut-in or when the flow rate has

stabilized, allowing the pressure to propagate radially through the reservoir.

3. Apply the radial flow model equations: The radial flow model equations are applied to the pressure data to estimate reservoir properties. The most common equation used is the radial diffusivity equation, also known as the "drawdown equation." It relates the pressure change over time to reservoir properties such as permeability, porosity, and reservoir thickness.

4. Analyze pressure responses: The pressure responses obtained from the radial flow model are compared to the actual pressure data. The analysis involves matching the modeled pressure response to the observed pressure data by adjusting the reservoir properties.

5. Derive reservoir properties: By adjusting the reservoir properties in the radial flow model, reservoir engineers can derive estimates of important parameters such as permeability, porosity, and skin factor. These estimates provide insights into the reservoir's productivity, connectivity, and near-wellbore effects.

It is important to note that despite the widespread application of the radial flow model, there are other, more complicated models that can be used instead. These models are able to take into account additional complexities that can be found in reservoirs, such as anisotropy, heterogeneity, or boundaries. However, in order to produce more realistic representations of the behavior of reservoirs in the real world, these models typically demand a larger amount of data as well as additional computer resources.

# Build A Pressure History

In pressure transient analysis, one of the most basic steps is the creation of a pressure history. It entails categorizing the pressure data that was acquired during a well test to establish a chronological record of pressure measurements at various sites inside the reservoir. This is done in order to analyze the data. This pressure history offers a rich dataset that can be utilized for further investigation and interpretation.

The process of building a pressure history involves the following steps:

1. Collect pressure measurements: During a well test, pressure measurements are taken at various points of interest within the reservoir. These measurements can be obtained using downhole pressure gauges, surface pressure gauges, or a combination of both. The measurements are typically recorded at regular time intervals or at specific points during the test.

2. Quality control and data validation: The collected pressure data undergoes quality control procedures to ensure its reliability and accuracy. This includes checking for sensor drift, calibration issues, and any anomalies or outliers in the data. Data validation techniques are applied to identify and address any errors or inconsistencies in the pressure measurements.

3. Data alignment: In some cases, pressure measurements from different gauges or sensors may need to be aligned to a common reference time. This alignment is necessary when pressure measurements are taken at different times or when multiple gauges are used to monitor pressure at various locations. Aligning the data ensures that pressure measurements correspond to the same time

reference, enabling meaningful comparisons and analysis.

4. Time synchronization: Pressure data from different gauges or sensors may have slight time discrepancies due to factors such as data transmission delays or differences in recording frequencies. Time synchronization techniques are applied to accurately align the pressure measurements in time, enabling a coherent and consistent pressure history.

5. Data interpolation: In situations where pressure measurements are not available at regular time intervals, data interpolation may be performed. This involves estimating pressure values at desired time points based on the available measurements. Interpolation techniques, such as linear interpolation or curve fitting methods, are used to fill in the gaps and create a continuous pressure history.

6. Pressure normalization: To facilitate comparison and analysis, pressure data may undergo normalization. This involves adjusting the pressure values to a common reference point or standard condition. Common normalization techniques include converting pressures to absolute or gauge pressures, adjusting for temperature variations, or referencing pressures to a specific datum.

7. Pressure history representation: Once the pressure data has been processed, validated, aligned, synchronized, and normalized, it is organized to create a pressure history. The pressure history typically consists of a time series dataset, where each entry corresponds to a specific location and time point, along with the corresponding pressure measurement.

In pressure transient analysis, the pressure history is used as a foundation for the further analysis that is performed. Engineers are able to examine pressure responses, determine flow regimes, and derive reservoir features using mathematical modeling and analysis because it gives a full record of pressure behavior through time at various points within the reservoir.

For the purpose of acquiring reliable and useful insights into reservoir behavior, maximizing well performance, and making informed decisions regarding reservoir management techniques, it is essential to develop pressure histories that are accurate and as detailed as possible.

## Identify the radial flow period

In pressure transient analysis, determining the radial flow period is an essential step because it establishes the portion of the pressure data that can be analyzed effectively with the help of the radial flow model. This portion of the data is determined by the portion of the pressure data that is suitable for analysis. The radial flow period is the time period during which the primary flow behavior in the reservoir is radial. This means that fluid flow happens primarily in a radial direction away from the wellbore during this time period, and this time period is referred to as the radial flow period.

The radial flow period normally takes place after the well has been shut in, which means that the flow of fluids into or out of the well has been halted. Radial flow refers to the flow of fluids in a circular pattern around the well. When the well is plugged, the pressure within the reservoir is able to rise and remain stable, which produces the circumstances essential for radial flow to occur. Throughout the reservoir, the pressure will gradually return to its original state as the shut-in period progresses, and fluid flow will radiate outward from the wellbore in all directions.

In order to determine the radial flow period, the data on the well's pressure that was gathered throughout the test is scrutinized and evaluated in great detail. When determining whether or not radial flow behavior is dominant, several methods and criteria are utilized, including the following:

1. Pressure stabilization: The pressure data is evaluated to observe when the pressure stabilizes. This occurs when the pressure readings exhibit minimal or negligible changes over time, indicating that the reservoir has reached a state of equilibrium.

2. Flow regime analysis: The pressure data is analyzed to identify any distinctive flow regimes. Initially, after shut-in, the pressure response may exhibit an early-time behavior known as the "wellbore storage" or "wellbore dominated" period. During this period, the pressure is mainly influenced by the wellbore itself, rather than the radial flow in the reservoir. Once the wellbore storage effects diminish, the pressure response transitions into the radial flow period.

3. Diagnostic plots: Diagnostic plots, such as pressure derivative plots, are examined to detect characteristic signatures of radial flow behavior. These plots provide valuable insights into the flow regime by highlighting distinct trends and patterns associated with radial flow.

4. Rate transient analysis: If flow rates are available, rate transient analysis can be performed to observe the stabilization of flow rates and the transition to radial flow behavior. When the flow rates have reached a steady state or stabilized condition, it indicates that radial flow dominates.

It is essential to keep in mind that the length of time for the radial flow period can change depending on the properties of the reservoir, the configuration of the well, and the testing conditions. In certain situations, the radial flow phase may only last for a brief amount of time, while in others, it may continue for a very longer period of time.

Following the determination of the radial flow period, the pressure data that falls within that period is chosen for additional investigation utilizing the radial flow model. The resulting estimates of reservoir parameters, such as permeability, porosity, and skin factor, offer invaluable insight into the behavior and performance of reservoirs. These estimates are derived from the data provided.

In conclusion, determining the radial flow period requires looking at the data on the pressure that was gathered during the well test. This is done in order to find the time period during which the radial flow behavior is most prevalent in the reservoir. This time period takes place after the well has been shut in or when flow rates have become stable, whichever comes first, and it enables the pressure to spread radially throughout the reservoir. Estimating the parameters of the reservoir and determining how well it is performing can both be accomplished through the use of the radial flow model, which can be made more accurate by picking the relevant data from within the radial flow period.

## Apply the radial flow model equations

Using the radial flow model equations, more especially the radial diffusivity equation or the drawdown equation, is a crucial stage in the process of performing pressure transient analysis in order to determine significant reservoir parameters. The radial diffusivity equation establishes a connection between the fluctuating pressure over time and several properties of the reservoir, such as the permeability, porosity, and thickness of the reservoir.

The concepts of fluid flow in porous medium are used to construct the equation for radial diffusivity. This equation is based on the assumption that radial flow occurs in a reservoir that is homogeneous and isotropic. The equation can be represented in a variety of different ways; however, the

39

pressure derivative form, also known as the Horner plot or log-log plot, is a form that is utilized frequently. This form makes it easier to analyze and interpret pressure transient data, and it does so in a more straightforward manner.

In its form that takes into account the pressure derivative, the equation for radial diffusivity is commonly stated as follows:

$$dp/dt = C * (\Delta p / \Delta t) / (t * r^2 * \mu * ct)$$

Where:

- dp/dt is the pressure derivative with respect to time (a dimensionless parameter).

- C is a dimensionless constant that depends on the wellbore and reservoir boundary conditions.

- $\Delta p$ / $\Delta t$ represents the change in pressure over time (pressure gradient).

- t is the time since the start of the well test.

- r is the radial distance from the wellbore to the point of interest in the reservoir.

- $\mu$ is the viscosity of the fluid.

- ct is the total compressibility of the reservoir rock and fluid system.

When you use the equation for radial diffusivity, you will find that in order to compute the pressure derivative, you will need to use the pressure data that was gathered during the well test. To calculate the pressure derivative, one must first find the derivative of the pressure data with respect to the logarithmic time scale. This can be accomplished through the use of numerical methods or specialized software.

Engineers are able to notice diverse flow regimes and discover unique patterns that provide insights into reservoir qualities if they plot the pressure derivative (dp/dt) vs logarithmic time on a log-log scale. This allows for the engineers to better understand reservoir properties. These patterns are as follows: 1

    1. Radial flow regime (early-time response): In the early stages of a well test, when the pressure disturbance is still propagating radially through the reservoir, the pressure derivative curve exhibits a

41

straight line with a slope of -0.5. This linear portion represents the radial flow regime and is used to estimate the reservoir's permeability and porosity.

2. Transition regime (intermediate-time response): As time progresses, the pressure derivative curve transitions from the straight line of the radial flow regime to a curved response. This transition indicates the start of wellbore storage and the influence of reservoir boundaries. The behavior of the pressure derivative curve in this regime helps to identify the boundaries and the presence of any reservoir heterogeneity.

3. Pseudo-steady-state regime (late-time response): In the later stages of the well test, when the pressure disturbance has propagated throughout the reservoir, the pressure derivative curve becomes relatively flat with a slope close to zero. This regime is known as pseudo-steady-state and is used to estimate the reservoir's thickness and to assess any near-wellbore effects such as skin or wellbore storage.

By analyzing the shape and behavior of the pressure derivative curve, engineers can estimate reservoir properties. For example:

- The slope of the linear portion during the radial flow regime provides an estimate of the permeability and porosity of the reservoir.

- The transition regime can provide information about reservoir boundaries, heterogeneity, and connectivity.

- The late-time behavior in the pseudo-steady-state regime allows for the estimation of the reservoir thickness and identification of near-wellbore effects.

The use of the radial diffusivity equation relies on a number of simplifying assumptions, the most essential of which is that the reservoir in question is both homogeneous and isotropic. It is imperative that this fact be brought to your attention. These assumptions might not hold true in real-world reservoirs, and it could be necessary to use more complicated models or make revisions in order to take into account variances in permeability, anisotropy, or any number of other reservoir complications.

In conclusion, the radial flow model equations, specifically the radial diffusivity equation, play a significant role in pressure transient analysis. This is especially true of the radial diffusivity equation. They provide a mathematical framework that may be used to analyze data on the pressure and estimate the parameters of the reservoir. Reservoir engineers are able to estimate permeability, porosity, and reservoir thickness by examining the pressure derivative curve. This analysis also helps them identify boundaries and near-wellbore effects, which are all helpful for attempts to optimize production and characterize reservoirs.

# Analyze pressure responses

The pressure data acquired from a well test is compared to the pressure responses predicted by the radial flow model as part of the pressure transient analysis process. The examination of pressure responses is a vital stage in the pressure transient analysis process. This analysis tries to match the modeled pressure response with the observed pressure data by modifying the features of the reservoir. This will lead to the calculation of reservoir parameters as well as an evaluation of the performance of both the well and the reservoir.

Graphical analysis, such as pressure derivative plots or type curve analysis, is commonly utilized in order to carry out the comparison between the modeled and observed pressure responses. The rate of change in pressure over a certain period of time is referred to as the pressure derivative, and it offers a wealth of information regarding the operation of the reservoir. Reservoir engineers are able to get insights into the parameters of the reservoir as well as discover any variations that may exist between the model and the real data if they conduct an analysis of the shape and characteristics of the pressure derivative curves.

Over the course of the analysis, the following procedures are normally carried out:

1. Convert pressure data to pressure derivatives:

Converting pressure data to pressure derivatives is a crucial step in pressure transient analysis. This transformation allows for a more detailed examination of the pressure response and helps identify distinct features that provide valuable information about the reservoir properties and flow behavior. By analyzing the pressure derivatives, reservoir engineers can gain insights into the reservoir's characteristics, such as boundaries, heterogeneities, and wellbore effects.

44

Differentiating the pressure data with regard to the passage of time is an essential step in the process of transforming raw pressure readings into pressure derivatives. The pressure derivative not only provides a more accurate depiction of the pressure's behavior across a variety of flow regimes, but it also reflects the rate of change in pressure in relation to the passage of time.

The pressure derivative transformation contributes in a number of ways to the improvement of the interpretation process:

> 1. Identification of characteristic slopes:
> Identification of characteristic slopes in the pressure derivative plot is a fundamental aspect of pressure transient analysis. These slopes provide valuable information about the flow regimes occurring within the reservoir, allowing reservoir engineers to infer important reservoir properties and behavior. The analysis of characteristic slopes aids in understanding the flow geometry and optimizing production strategies.

The -1/2 slope is a distinctive slope that may be seen in the pressure derivative plot when radial flow is occurring. It is one of the most well-known and extensively utilized characteristic slopes. As fluids flow radially away from the wellbore and into the surrounding reservoir, this is an example of radial flow. A characteristic straight line can be seen in the pressure derivative map.

This line has a slope of -1/2, and it has important repercussions for the parameters of the reservoir.

The slope of -1/2 is the one that corresponds to the traditional behavior of radial flow, which is defined by Darcy's law and the diffusivity equation. Within the parameters of this flow regime, the response of the pressure is proportional to the cube root of the passage of time. It may be deduced from the presence of this characteristic slope that the reservoir is acting as a homogeneous and isotropic medium. This means that there are no substantial barriers, heterogeneities, or wellbore influences influencing the flow of fluid within the reservoir.

The -1/2 slope is significant because of its link with a number of different reservoir features, including the following:

1. Permeability: The -1/2 slope is primarily sensitive to the permeability of the reservoir. As fluids flow through the reservoir, the rate of pressure decline is proportional to the square root of time, which is influenced by the permeability. A steeper -1/2 slope indicates higher permeability, while a shallower slope suggests lower permeability.

2. Drainage area: The magnitude of the -1/2 slope is related to the size of the drainage area contributing to the pressure response. A larger drainage area results in a more pronounced -1/2 slope, indicating a greater extent of reservoir depletion and connectivity to the wellbore.

3. Reservoir boundaries: The absence of significant deviations from the -1/2 slope suggests that the reservoir is not influenced by boundaries or barriers that can affect fluid flow. If deviations from the -1/2 slope are observed, it may indicate

the presence of boundaries or reservoir compartments that restrict the radial flow behavior.

4. Skin factor: The -1/2 slope can also provide insights into the presence of near-wellbore effects, such as wellbore storage or skin damage. If the pressure derivative plot deviates from the expected -1/2 slope, it may suggest the existence of wellbore effects that affect the pressure response and require further investigation.

Although the slope of -1/2 is the most well-known characteristic slope, other slopes can occur in the pressure derivative plot to indicate different flow regimes or reservoir complexities. While the slope of -1/2 is the most well-known distinctive slope, other slopes can emerge. For instance, when the flow is linear, and the reservoir acts as though it were infinite and homogenous, the pressure derivative plot will have a slope of -1. The presence of limits, fractures, or heterogeneities that alter fluid flow can be indicated by the presence of other slopes.

The identification and interpretation of characteristic slopes should be done in conjunction with other information and data that is available, such as well logs, production history, and knowledge of the geological features that are present in the reservoir. This is an important point to keep in mind because it is necessary to do so. Validating the observed slopes and improving the calculation of reservoir attributes can both be helped by conducting an exhaustive examination.

In a nutshell, the identification of characteristic slopes in the pressure derivative plot, such as the -1/2 slope, is essential for the understanding of pressure transients. Insights regarding flow regimes, permeability, drainage area, reservoir boundaries, and wellbore effects can be gained from these slopes. Reservoir engineers are able to make educated decisions on the characterization of reservoirs, the optimization of production, and the evaluation of well performance if they conduct an analysis of the pressure derivative plot and the slopes associated with it.

2. Detection of inflection points:
Detecting inflection points in the pressure derivative plot is a crucial aspect of pressure transient analysis. Inflection points are significant features that indicate transitions between different flow regimes within the reservoir. They provide valuable information about reservoir boundaries, heterogeneities, and near-wellbore effects that impact the pressure behavior.

Changes in the gradient of the pressure derivative curve might be interpreted as the presence of inflection points. These inflection points can reveal information about the behavior and qualities of the reservoir if you pay attention to their shape and attributes. In the context of pressure transient analysis, the following is an in-depth discussion of the significance of inflection points as well as their interpretation:

1. Identification of flow regime transitions:
Inflection points represent the transitions between different flow regimes within the reservoir. For example, during the early stages of a well test, the pressure derivative curve typically exhibits a straight line with a slope of -1/2, indicating radial flow. As time progresses, the flow behavior may

change due to reservoir heterogeneities, boundaries, or wellbore effects. The detection of inflection points helps identify these transitions and provides information about the presence of different flow regimes such as linear flow, bilinear flow, or boundary-dominated flow.

2. Insight into reservoir boundaries: Inflection points in the pressure derivative plot can indicate the presence of reservoir boundaries or barriers that influence fluid flow. A significant change in the slope of the pressure derivative curve suggests that the flow behavior is affected by a boundary or barrier, causing a shift in the flow regime. The inflection points provide information about the distance from the wellbore to the reservoir boundaries, enabling engineers to assess reservoir geometry and connectivity.

3. Assessment of reservoir heterogeneities: Inflection points can also signify the presence of reservoir heterogeneities. Variations in rock properties, such as permeability or porosity, can lead to changes in flow behavior. Inflection points in the pressure derivative curve may indicate the transition between regions with different permeability values or the presence of layers with contrasting properties. By analyzing these inflection points, reservoir engineers can gain insights into the spatial distribution of heterogeneities and their impact on fluid flow within the reservoir.

4. Evaluation of near-wellbore effects: Near-wellbore effects, such as wellbore storage or formation damage, can significantly influence pressure behavior. Inflection points in the pressure derivative curve can indicate the presence of near-wellbore effects and their impact on flow behavior. Deviations from the expected flow regime can provide information about the degree and type of near-wellbore damage, allowing engineers to assess its effect on well performance and make decisions regarding remedial actions.

5. Quantification of reservoir parameters: Inflection points in the pressure derivative curve provide a basis for estimating reservoir parameters. The shape and position of these inflection points, along with other information obtained from the pressure transient analysis, can be used to estimate important reservoir properties, such as reservoir thickness, permeability, porosity, and skin factor. These estimates contribute to a better understanding of the reservoir's characteristics and aid in optimizing production strategies.

In conclusion, recognizing inflection points in the pressure derivative plot is essential for comprehending shifts in flow regimes, locating reservoir limits and heterogeneities, gauging the effects of the wellbore and reservoir proximity, and measuring reservoir attributes. The study of inflection points helps to improve the understanding of pressure transient data and gives useful information that may be used for the characterization of reservoirs and the optimization of production.

3. Differentiation of noise and signal:

Differentiating noise from the signal is an important aspect of pressure transient analysis. Pressure data acquired during well tests can be affected by various sources of noise, such as measurement errors, instrumentation limitations, wellbore effects, or external disturbances. These noise components can obscure the underlying reservoir behavior, making it challenging to extract meaningful information and interpret the pressure response accurately. Converting the pressure data to pressure derivatives helps differentiate noise from the signal, enhancing the interpretation process and facilitating a clearer understanding of the reservoir behavior.

Recognizing inflection points in the pressure derivative plot is essential for a number of reasons, including but not limited to: understanding shifts in flow regimes; locating reservoir limits and heterogeneities; gauging the effects of the wellbore and reservoir proximity; measuring reservoir attributes; and locating reservoir limits and heterogeneities. The study of inflection points helps to improve the understanding of pressure transient data and gives useful information that may be used for the characterization of reservoirs and the optimization of production. The study of inflection points also contributes to the improvement of reservoir characterization.

1. Enhanced signal-to-noise ratio: The differentiation process helps improve the signal-to-noise ratio of the pressure data. By emphasizing the dynamic changes in pressure, the noise components, which are typically slower and less

significant, become relatively smaller in magnitude compared to the amplified signal. This improvement in the signal-to-noise ratio allows for a clearer identification of meaningful features in the pressure response.

2. Highlighting transient behaviors: Pressure transient analysis focuses on capturing transient behaviors and deviations from steady-state conditions. The noise components in the pressure data, being more random and less correlated with the transient phenomena of interest, tend to average out when differentiating. As a result, the pressure derivatives emphasize the transient behaviors, such as flow regime transitions, pressure buildup or drawdown events, and characteristic slopes associated with specific reservoir properties.

3. Noise filtering and denoising: The differentiation process acts as a noise filtering technique. The noise components, often manifested as high-frequency oscillations or random variations in the pressure data, tend to have a higher frequency content compared to the meaningful signal. Through differentiation, these high-frequency noise components are accentuated, while the lower-frequency signal is preserved. This denoising effect aids in removing unwanted noise and extracting the essential features of the pressure response.

4. Identification of distinct features: The differentiation of pressure data to pressure derivatives can help identify distinct features that correspond to specific reservoir properties or flow behavior. The noise components, being uncorrelated with the reservoir characteristics, typically result in random fluctuations in the pressure derivatives. In contrast, the meaningful features associated with the reservoir properties, such as characteristic slopes or inflection points, tend to exhibit clear patterns in the pressure derivatives. By focusing on these distinct features, reservoir engineers can gain insights into the reservoir's behavior and properties.

5. Improved interpretation and model calibration: The noise differentiation process contributes to a clearer interpretation of the pressure response and facilitates more accurate model calibration. The amplified signal in the pressure derivatives enables a more precise identification of flow regime boundaries, estimation of reservoir parameters, and assessment of well performance. By analyzing the denoised pressure derivatives, engineers can refine the reservoir model and optimize the model parameters to achieve a better match between the modeled and observed data.

In conclusion, one of the most important aspects of pressure transient analysis is the separation of noise from the signal, which is accomplished by the conversion of pressure data to pressure derivatives. It improves the signal-to-noise ratio, draws attention to transient behaviors, filters out noise, finds distinguishable characteristics, and makes interpretation and model calibration easier. Reservoir engineers are able to gain a better understanding of the behavior of the reservoir, make more accurate estimates of the properties of the reservoir, and make more informed decisions regarding reservoir management and production optimization when they place more emphasis on the essential features of the pressure response.

4. Sensitivity to specific reservoir properties:

The sensitivity of pressure derivative plots to specific reservoir properties is a key aspect of pressure transient analysis. By examining the shape and characteristics of the pressure derivative curve, reservoir engineers can gain valuable insights into the presence of reservoir boundaries, barriers, and compartmentalization. This information is crucial for understanding the reservoir's behavior, connectivity, and potential production challenges.

The pressure derivative plot exhibits distinct patterns that are sensitive to various reservoir properties, including:

1. Reservoir boundaries: Reservoir boundaries, such as faults or geological formations, can significantly impact fluid flow within the reservoir. The pressure derivative plot can provide indications of the presence and distance to these boundaries. In the early-time portion of the plot, the slope or curvature can reveal the proximity of boundaries. Steeper slopes may indicate closer boundaries, while flatter slopes suggest a larger distance to the boundaries.

2. Barriers and compartments: Reservoir heterogeneities or barriers, such as high-permeability streaks or faults that impede fluid flow, can create compartmentalization within the reservoir. The pressure derivative plot can help identify these compartments by exhibiting distinct behaviors between different parts of the reservoir. Sharp changes or discontinuities in the pressure derivative curve can indicate the presence of compartments with different flow characteristics.

3. Faults and fractures: Faults and fractures in the reservoir can significantly affect fluid flow behavior. These features can act as conduits or barriers to fluid movement, leading to localized pressure changes. The pressure derivative plot can exhibit distinctive signatures, such as step-like changes or abrupt shifts, indicating the influence of faults or fractures on the reservoir behavior.

4. Boundaries with other formations: The pressure derivative plot can provide insights into the interactions between the reservoir and surrounding formations. For example, the plot may exhibit a specific response when the reservoir is in contact with a different rock type or when fluid flow occurs across interfaces. These responses can help in understanding the connectivity and fluid movement between different formations.

It is essential to keep in mind that the interpretation of pressure derivative charts calls for both specialized knowledge and careful analysis of the other data and information that is currently available about the reservoir. In order to provide meaningful interpretations, the patterns that are seen while looking at the pressure derivative plot are compared to known geological and reservoir properties.

In addition to the visual interpretation, quantitative analysis methods such as type curve matching can be utilized to match the observed pressure derivative plots with pre-established type curves that represent a variety of reservoir conditions and properties. This is accomplished by comparing the plots of the pressure derivative with the pre-established type curves. Because of this procedure, engineers are able to estimate crucial factors like the limits of reservoirs, the sizes of compartments, and the differences in permeability.

Reservoir engineers can get significant insights into the behavior and connectivity of a reservoir by making use of the sensitivity of pressure derivative plots to specific reservoir parameters. This information is essential for making educated decisions on the optimization of hydrocarbon recovery, production strategies, and the development of reservoirs.

5. Model calibration and parameter estimation:

Model calibration and parameter estimation play a crucial role in pressure transient analysis. This process involves comparing the shape and features of the pressure derivative curve obtained from the actual pressure data to those generated by the radial flow model. By adjusting the model parameters iteratively, reservoir engineers can achieve the best match between the modeled and observed pressure data. This iterative calibration process allows for the estimation and refinement of key reservoir properties, such as permeability, porosity, and skin factor.

The calibration and parameter estimation process typically involve the following steps:

1. Initial estimation: The pressure transient analysis starts with an initial estimation of the reservoir properties. These initial estimates are based on available geological and engineering data, historical production information, and prior knowledge of the reservoir.

2. Generate model pressure responses: Using the initial reservoir property estimates, the radial flow model is employed to generate modeled pressure responses. The model incorporates assumptions about the reservoir geometry, fluid properties, and wellbore conditions.

3. Compare modeled and observed pressure derivatives: The modeled pressure responses are transformed into pressure derivatives, and the shape and features of the pressure derivative curve are compared to the observed pressure derivative curve. The goal is to identify similarities and differences between the two curves.

4. Adjust model parameters: To improve the match between the modeled and observed pressure derivatives, the model parameters are adjusted. The parameters that can be modified include permeability, porosity, reservoir thickness, skin factor, and sometimes boundary conditions or wellbore storage effects. This adjustment process is often performed using optimization algorithms that seek to minimize the differences between the modeled and observed pressure derivatives.

5. Iterate and refine: The process of adjusting model parameters and comparing the modeled and observed pressure derivatives is iterative. Reservoir engineers make incremental changes to the model parameters, re-generate the modeled pressure responses, transform them into pressure derivatives, and compare them to the observed data. This iterative process is repeated until a satisfactory match between the modeled and observed pressure derivatives is achieved.

6. Evaluate uncertainty and sensitivity: As part of the calibration process, it is important to assess the uncertainty and sensitivity of the estimated reservoir properties. This involves quantifying the range of possible values for the parameters and evaluating their impact on the match between the modeled and observed pressure derivatives. Sensitivity analysis helps in identifying the parameters that have the most significant influence on the pressure response and allows for

a better understanding of the uncertainties associated with the estimated reservoir properties.

7. Final estimation and interpretation: Once a satisfactory match between the modeled and observed pressure derivatives is obtained, the final estimates of the reservoir properties, such as permeability, porosity, and skin factor, are determined. These estimates are used to characterize the reservoir and evaluate its performance. Interpretation of the estimated parameters provides insights into the reservoir's flow behavior, connectivity, and near-wellbore effects.

It is essential to keep in mind that the process of parameter estimation and calibration demands a high level of competence and experience. Engineers working on reservoirs need to have a thorough understanding of both the reservoir and its behavior in order to make educated decisions regarding the change of model parameters. In addition, the process may involve the application of modern optimization strategies as well as statistical analysis in order to guarantee accurate and trustworthy parameter estimate.

In a nutshell, the process of model calibration and parameter estimation in pressure transient analysis involves comparing the shape and features of the pressure derivative curve obtained from the actual pressure data to those generated by the radial flow model. This is done in order to determine whether or not the radial flow model is accurate. Reservoir engineers can improve the accuracy of their estimations of reservoir attributes by iteratively modifying the parameters of the model. This procedure permits the assessment of critical characteristics like as permeability, porosity, and skin factor, which provides helpful insights into the behavior of the reservoir and optimizes the performance of both the well and the reservoir.

59

In conclusion, one of the most important steps in the process of pressure transient analysis is transforming pressure data to pressure derivatives. It does this by highlighting identifiable features, such as typical slopes and inflection points, that correspond to specific reservoir parameters and flow behavior. This makes the interpretation process easier. The pressure derivative transformation is useful for determining flow regimes, locating reservoir limits, distinguishing noise from the signal, increasing sensitivity to particular reservoir features, and making model calibration and parameter estimates easier. Reservoir engineers can get significant insights into the features of the reservoir and be in a better position to make decisions regarding reservoir management and production optimization if they conduct an analysis of the pressure derivatives.

## Identify key pressure derivative signatures

In pressure transient analysis, one of the most important aspects is the identification of key pressure derivative fingerprints. The pressure derivative plots can reveal a wealth of information regarding the parameters of the reservoir as a result of the characteristic patterns that are produced by the various flow regimes and reservoir properties. Reservoir engineers are able to deduce information about the features of the reservoir, as well as its boundaries, heterogeneities, and effects near the wellbore, by detecting and interpreting these signatures.

On the pressure derivative plot, the signatures of the pressure derivative can be seen as distinct patterns or forms. When attempting to decipher these signatures, it is necessary to take into account their gradients, forms, and connections to the passage of time. Some c

1. Straight lines with a slope of -1/2: This is a characteristic signature associated with radial flow. In a homogeneous and isotropic reservoir, during the radial flow regime, the pressure derivative plot exhibits a straight line with a slope of -1/2. This

slope represents the decline of pressure as inversely proportional to the square root of time. The presence of this signature indicates a well-connected reservoir with a uniform permeability distribution.

2. Curved lines or multiple straight lines: These signatures often indicate the presence of reservoir heterogeneities or boundaries. Curved lines with different slopes or multiple straight lines with different slopes suggest the occurrence of different flow regimes or the influence of reservoir compartments. These signatures provide valuable insights into reservoir compartmentalization, the presence of barriers, or variations in permeability within the reservoir.

3. Early time behavior and wellbore storage: During the early time period of a well test, the pressure derivative plot may exhibit distinct signatures that indicate wellbore storage effects. These signatures appear as curved lines with positive slopes and are influenced by the volume of fluid stored within the wellbore. Analyzing these signatures helps in quantifying wellbore storage effects and separating them from the reservoir response.

4. Late time behavior and boundary-dominated flow: As the well test progresses, the pressure derivative plot may exhibit signatures that indicate boundary-dominated flow. Boundary-dominated flow occurs when the influence of reservoir boundaries becomes significant. This signature is observed as a change in the slope of the pressure derivative curve or as a flattening of the curve. It provides insights into the presence and proximity of reservoir boundaries, which affect fluid flow and reservoir performance.

5. Skin effect and wellbore damage: The presence of a skin effect,

which represents near-wellbore damage or stimulation effects, can be inferred from the pressure derivative plot. It appears as a deviation from the expected pressure derivative behavior. A positive skin effect manifests as a downward shift in the pressure derivative curve, indicating wellbore damage or reduced near-wellbore permeability. Conversely, a negative skin effect suggests wellbore stimulation or increased near-wellbore permeability.

6. Intermediate slopes or shapes: The pressure derivative plot may exhibit intermediate slopes or complex shapes that do not correspond to the idealized signatures mentioned above. These patterns often arise due to reservoir complexities, such as anisotropy, heterogeneity, or non-uniform permeability distributions. Analyzing these intermediate signatures requires additional considerations, such as applying more complex mathematical models or considering the influence of reservoir layers or fractures.

Reservoir engineers are able to obtain vital insights into the properties of the reservoir by recognizing these key pressure derivative signatures and understanding the interpretations of those signatures. This information is useful for predicting permeability, locating the boundaries of the reservoir, quantifying the effects of the wellbore, assessing the reservoir's heterogeneities, and developing production methods for maximum efficiency. In addition to this, it assists in the process of making educated judgments on reservoir management, well completion designs, and enhanced oil recovery methods.

## Match modeled pressure responses to observed data

In pressure transient analysis, one of the most important steps is getting the modeled pressure responses to match up with the actual data. During this stage of the process, the radial flow model is used to generate simulated pressure responses depending on the properties of the reservoir that are expected to exist. After then, the simulated pressure responses are tweaked repeatedly by changing important parameters such as permeability, porosity, reservoir borders, and skin factor. This process continues until the simulated pressure responses have the best possible match with the actual pressure data.

These are the actions that must be taken in order to successfully match the modeled pressure responses to the observed data:

1. Initial parameter estimation: Initially, reservoir engineers make an educated guess or estimate for the reservoir properties based on available information, such as geological data, prior knowledge of similar reservoirs, or nearby well data. These estimated values are used as the starting point for the modeling process.

2. Simulate pressure responses: Using the assumed reservoir properties, the radial flow model is utilized to generate simulated pressure responses. The model calculates the expected pressure behavior over time based on the assumptions of fluid flow occurring radially from the wellbore into the reservoir.

3. Compare with observed data: The simulated pressure responses are then compared to the observed pressure data collected during the well test. The comparison is typically done graphically by overlaying the modeled and observed pressure data on a pressure versus time plot.

4. Adjust reservoir properties: Based on the comparison between the simulated and observed pressure responses, adjustments are

made to the reservoir properties. The parameters that are commonly adjusted include permeability, porosity, reservoir boundaries, and skin factor. These adjustments aim to improve the match between the modeled and observed pressure data.

5. Iterative refinement: The process of adjusting the reservoir properties and comparing the simulated and observed pressure responses is repeated iteratively. With each iteration, the values of the reservoir parameters are refined to converge towards a better match with the observed pressure data. This iterative refinement process continues until an acceptable match is achieved, indicating that the simulated pressure responses adequately represent the actual reservoir behavior.

6. Sensitivity analysis: During the iterative process, sensitivity analysis may be performed to assess the influence of different reservoir parameters on the pressure responses. By systematically varying one parameter while keeping others constant, the impact of each parameter on the match between the modeled and observed data can be evaluated. This analysis helps identify the most influential parameters and provides insights into the reservoir behavior.

7. Uncertainty assessment: It is important to consider the uncertainties associated with the estimated reservoir properties. Uncertainty analysis techniques, such as Monte Carlo simulations or sensitivity analysis with multiple parameter sets, can be employed to assess the range of possible reservoir property values that still provide a reasonable match to the observed data. This helps quantify the confidence level in the estimated reservoir properties and their impact on the pressure responses.

Reservoir engineers are able to improve their estimation of reservoir properties if they are successful in properly matching the modeled pressure responses to the observed data. The improved features, such as permeability and skin factor, provide vital insights into the productivity of the reservoir as well as its connectivity and effects around the wellbore. It is essential to have this knowledge in order to make educated choices regarding the management of reservoirs, the optimization of production, and the economic viability of the well or field.

## Interpret reservoir properties

The pressure transient analysis process is not complete without the interpretation of the reservoir properties. The modified reservoir attributes can be calculated once a good match has been made between the modeled pressure responses and the observed pressure data. This provides vital insights into the features and performance of the reservoir. These variables, which play a substantial part in defining the reservoir's production potential, connectivity, and near-wellbore effects, include permeability, porosity, reservoir thickness, and skin factor. Other properties that play a role include reservoir thickness and skin factor.

1.  Permeability: The ease with which fluids are able to flow through the rock formation is referred to as the rock formation's permeability, which is an essential reservoir feature. Estimating the reservoir's permeability is done as part of the pressure transient analysis process by using the matched pressure responses derived from the radial flow model. The permeability estimation contributes to the evaluation of the reservoir's capacity to transport fluids and has an impact on production rates, well performance, and management strategies for reservoirs.

2.  Porosity: Porosity is the volume of empty space that exists inside the reservoir rock and has the capability of holding fluids.

Reservoir engineers are able to make an educated guess about the effective porosity of the reservoir by analyzing the matched pressure responses. It is vital to calculate effective porosity in order to determine the overall amount of hydrocarbons that a reservoir is capable of holding as well as the flow capacity of the rock. Effective porosity accounts for the interconnected void space that contributes to fluid flow.

3. Reservoir Thickness: The examination of pressure responses can yield information that is helpful in determining the thickness of the reservoir. Reservoir engineers are able to determine the thickness of the reservoir by changing the radial flow model in such a way that it matches the observed pressure data. The behavior of fluid flow, volumetric calculations, and the total hydrocarbon potential of the reservoir are all impacted by the thickness of the reservoir.

4. Skin Factor: The skin factor is an important measure that quantifies the near-wellbore effects, such as formation damage or wellbore storage, which impact the flow of fluids from the reservoir into the wellbore. These near-wellbore effects can be impacted by a variety of factors. The skin factor can be calculated by pressure transient analysis by changing the radial flow model to match the observed pressure data. This is done in order to determine the skin factor. The presence of additional pressure losses or gains, as the case may be, near the wellbore is indicated by a skin factor that is either positive or negative.

Calculating these reservoir attributes through the use of pressure transient analysis offers very helpful insights into the features and performance of the reservoir, including the following:

a. Productivity Potential: When evaluating the production potential of a reservoir, it is necessary to have access to certain pieces of information, some of which can be derived from the reservoir's attributes, such as its permeability and porosity. They contribute to the calculation of the anticipated flow rates and production profiles, as well as the eventual recovery of hydrocarbons from the reservoir.

b. Connectivity: The evaluation of the connectivity of the reservoir is made possible for reservoir engineers through the examination of pressure responses. They are able to acquire insights regarding the connectivity of the reservoir with wells or compartments in the surrounding area if they examine the pressure data that have been matched. This knowledge is essential for constructing effective reservoir draining techniques, as well as optimizing well placement, planning well patterns, and designing well patterns.

c. Near-Wellbore Effects: The estimated skin factor sheds light on the near-wellbore influences that have an effect on well performance. When the skin values are negative, enhanced wellbore conditions are suggested, whereas positive skin values signal the possibility of formation damage or a reduction in well productivity. When making decisions on well stimulation, remedial treatments, and production optimization, having a thorough understanding of and being able to quantify the near-wellbore effects is absolutely necessary.

In a nutshell, evaluating reservoir attributes through pressure transient analysis includes comparing the modeled pressure responses to the actual pressure data in order to estimate permeability, porosity, reservoir thickness, and skin factor. These predicted features provide vital insights into the reservoir's potential for production, connectivity, and effects around the wellbore. This knowledge helps guide decisions about reservoir management and production optimization tactics, and it also improves our overall understanding of how reservoirs behave.

## Evaluate well and reservoir performance

Evaluation of the performance of both the well and the reservoir is an essential component of pressure transient analysis. The interpretation of the pressure responses that were gathered during the testing of the well offers extremely helpful insights into the operation of the well as well as the productivity of the reservoir. Reservoir engineers are able to evaluate the performance of the well and make educated judgments on reservoir management and production optimization by comparing the predicted reservoir parameters with preset criteria and industry benchmarks.

The following is a list of the most important factors to consider when doing pressure transient analysis to evaluate well and reservoir performance:

1. Performance comparison: The estimated reservoir properties, such as permeability, porosity, and skin factor, are compared to predefined criteria or industry standards. These criteria may include minimum acceptable permeability values, maximum allowable skin factors, or target porosity ranges. By evaluating the estimated reservoir properties against these benchmarks, engineers can determine whether the well is performing optimally or if further actions are necessary.

2. Productivity assessment: The analysis of pressure responses helps in assessing the productivity of the well. By estimating reservoir properties, engineers can evaluate the well's ability to deliver fluids from the reservoir to the wellbore. Parameters such as well productivity index (PI) and flow capacity can be derived from the pressure responses and compared to desired productivity levels. This assessment provides insights into the efficiency of fluid flow and aids in identifying potential production limitations.

3. Identification of formation damage: Pressure transient analysis can help identify the presence of formation damage near the wellbore. Formation damage refers to any impairment in reservoir permeability caused by factors such as drilling mud invasion, fluid filtrate invasion, or solids deposition. By comparing the estimated skin factor, which quantifies near-wellbore damage, with expected values, engineers can identify the extent and impact of formation damage on well performance. This information is crucial for determining if remedial actions, such as acidizing or hydraulic fracturing, are necessary to restore or enhance well productivity.

4. Detection of reservoir heterogeneities: Pressure transient analysis aids in identifying reservoir heterogeneities or variations in reservoir properties. These variations can occur due to differences in rock properties, fluid distribution, or geologic features. By comparing the modeled pressure responses to the observed data, engineers can detect deviations that may indicate the presence of reservoir heterogeneities. Understanding these variations helps in characterizing the reservoir and designing appropriate production strategies to optimize recovery.

5. Boundary identification: Pressure transient analysis provides insights into the presence and behavior of reservoir boundaries. Reservoir boundaries can influence fluid flow patterns and impact well performance. By examining the pressure responses, engineers can identify pressure signatures that indicate the proximity to reservoir boundaries or barriers. This knowledge helps in determining reservoir connectivity, understanding fluid drainage patterns, and making decisions regarding well spacing and reservoir management.

In general, doing pressure transient analysis while analyzing well and reservoir performance is essential for optimizing production and making informed decisions in reservoir engineering. This is because these two aspects are directly related.

Engineers are able to get significant insights into the behavior of the well and reservoir by comparing predicted reservoir parameters with preset criteria, estimating productivity, diagnosing formation damage, detecting reservoir heterogeneities, and defining boundaries. With this information, reservoir management techniques may be optimized, production optimization efforts can be strengthened, and maximum hydrocarbon recovery can be achieved.

It is essential to keep in mind that pressure transient analysis is an iterative process. Moreover, numerous models and methods may be utilized in order to increase the degree to which the modeled and observed pressure responses correspond to one another. They may include things like taking into account models with a higher level of complexity, including additional reservoir heterogeneities or anisotropies, or making use of advanced analytical techniques.

In a nutshell, the comparison of the modeled pressure responses produced by the radial flow model to the observed pressure data constitutes the bulk of the pressure transient analysis portion of the study of pressure responses. The results of this comparison make it possible to modify the parameters of the reservoir in order to reach a satisfactory fit. The reservoir parameters can be estimated, and both the well performance and the performance of the reservoir can be evaluated, through the process of analyzing the pressure derivative plots and matching the modeled data to the observed data. This research offers useful information that can be used for the characterisation of reservoirs, the optimization of production, and the making of decisions in reservoir engineering.

# Derive reservoir properties

The procedure of deducing the properties of the reservoir is an integral part of the pressure transient analysis method. This is as a result of the fact that it enables reservoir engineers to estimate important parameters that influence fluid flow and determine the behavior of the reservoir. This is the primary reason for this. Engineers are able to obtain estimates of important reservoir properties such as permeability, porosity, and skin factor by adjusting the reservoir properties in the radial flow model and comparing the modeled pressure responses to the observed pressure data. This allows the engineers to obtain estimates of important reservoir properties such as permeability, porosity, and skin factor. Because of this, the engineers are able to generate estimates of the reservoir parameters that are more precise.

The ability of a reservoir rock to allow fluids to pass through it is referred to as its permeability. It is an essential parameter that has a direct influence on the rates at which fluids flow and the overall productivity of the reservoir. Engineers make adjustments to the permeability of the radial flow model using pressure transient analysis in order to get the best match possible between the pressure responses that are modeled and those that are observed. It is vital to accurately anticipate production rates, design optimal well completions, and optimize reservoir development plans in order to make use of the obtained permeability value, which offers insights into the rock's ability to transport fluids and gives these insights.

Porosity is the volume of void space within the reservoir rock that is able to hold fluids, and it is referred to as "porosity." The overall amount of hydrocarbons that a reservoir is capable of holding as well as the flow capacity of the rock are both significantly impacted by this key feature. Adjusting the model and comparing the pressure responses are two of the steps involved in pressure transient analysis, which both help in calculating the effective porosity of the reservoir. This calculation gives significant information regarding the storage capacity of the reservoir as well as the possibility for fluid accumulation.

The skin factor is a dimensionless quantity that indicates the near-wellbore damage or stimulation effects in the reservoir. It is also known as the "skin factor." It provides a numerical representation of the variance in pressure near the wellbore in comparison to the behavior that would be expected from an idealized homogeneous and isotropic reservoir. When it comes to the reservoir that is located close to the wellbore, positive skin factors imply damage such as formation damage or wellbore storage effects, whilst negative skin factors suggest stimulation effects. Engineers are able to achieve a more realistic picture of the behavior of the reservoir by modifying the skin factor in the radial flow model. This adjustment allows the engineers to take into consideration the impact of near-wellbore effects.

Connectivity between reservoirs: In addition to this, pressure transient analysis can shed light on the connectedness of the reservoir. Connectivity refers to the channels or passageways that fluids use to move about inside the reservoir. Engineers are able to deduce the connectivity between various sections of the reservoir by modifying the model parameters and matching the pressure responses. The optimization of reservoir development plans, the design of well placement methods, and the determination of the efficacy of reservoir management techniques such as water flooding and hydraulic fracturing all require this information. It is vital.

Pressure transient analysis gives engineers the ability to examine and quantify the near-wellbore impacts, such as wellbore storage, formation damage, or stimulation effects. These types of effects can be caused by a variety of factors. The influence of these impacts can be taken into account in the study by making necessary adjustments to the model's parameters, most notably the skin factor. For the purposes of determining well performance, optimizing well completions, and identifying any necessary corrective actions to increase output, it is essential to understand and quantify these effects.

It is essential to keep in mind that the process of deriving reservoir parameters through pressure transient analysis is an iterative one. In order to get a good match between the modeled and observed pressure responses, engineers make adjustments to the model parameters. In particular, they focus on permeability, porosity, and the skin factor. The parameters that were extracted from the reservoir provide useful insights into the productivity, connectivity, near-wellbore effects, and overall performance of the reservoir. These insights help in the process of making educated judgments on the management of reservoirs, the optimization of production, and the maximization of hydrocarbon recovery.

# Complex models

The following are three models that are considered to be more complicated and are utilized in pressure transient analysis:

## Dual Porosity/Dual Permeability Model:

The dual porosity/dual permeability model is a more complex model that is utilized in pressure transient analysis. This type of model is particularly useful in reservoirs that have substantial variability and complex flow behavior. It takes into consideration the existence of two separate systems within the reservoir, which are known as the matrix system and the fracture system respectively. This model acknowledges that fluid flow happens in these two systems in a distinct manner and takes into account both the specific attributes of each system as well as the interactions between them.

Within the framework of the dual porosity/dual permeability paradigm, the matrix system is meant to stand in for the low-permeability rock matrix. This matrix serves as a medium for storing the fluids. On the other hand, the fractures create routes with high permeability, which makes it easier for fluid to move through the rock. Both the matrix and the fracture systems have their very own unique porosity and permeability values, respectively.

The model takes into account the various flow mechanisms and transfer rates that are present in the matrix as well as the fracture systems. It takes into account events like matrix imbibition and fluid exchange between the

73

matrix and fractures. Both of these events involve the matrix absorbing fluids and the fractures exchanging fluids. In most cases, transfer coefficients or exchange parameters are utilized in order to characterize the transfer rates that occur between the matrix and the fracture systems.

Consider a hypothetical fractured reservoir so that we can demonstrate how the dual porosity/dual permeability concept can be applied in a real-world setting. In this particular illustration, the reservoir is made up of low-permeability matrix rock that is peppered with discrete fractures that function as preferential flow pathways.

1. Reservoir Characterization:

The first step is to characterize the reservoir properties. This includes determining the matrix and fracture permeabilities, porosities, and other relevant parameters. Core data, well logs, and production data can be utilized to estimate these properties.

2. Model Formulation:

The dual porosity/dual permeability model is formulated by representing the matrix and fracture systems as separate regions with their respective properties. The governing equations for fluid flow, such as Darcy's law, are applied to each system. The transfer rates between the matrix and fracture systems are incorporated using appropriate transfer coefficients or exchange parameters.

3. Well Test Analysis:

During the well test analysis, pressure data is obtained

from pressure gauges or downhole tools. This data is then interpreted using the dual porosity/dual permeability model to estimate the reservoir properties and assess the performance.

4. Model Calibration:

The model parameters, including the transfer coefficients, are calibrated to match the observed pressure data. This involves adjusting the parameters iteratively until a satisfactory match between the modeled pressure response and the observed data is achieved.

5. Interpretation of Results:

Once the model is calibrated, the estimated reservoir properties can be used to evaluate the behavior of the fractured reservoir. This includes assessing the storage capacity and flow characteristics of the matrix and fractures, identifying any heterogeneities or barriers that influence fluid flow, and analyzing the connectivity between the matrix and fracture systems.

The dual porosity/dual permeability model provides a more accurate portrayal of the complex flow dynamics in fractured reservoirs than alternative models. Engineers are able to optimize production strategies and make educated decisions regarding reservoir management when they have a better understanding of the interaction that occurs between the matrix and the fracture systems. This is made possible by a better comprehension of the dynamic that exists between the two.

It is essential to point out that the implementation of the dual porosity/dual permeability model necessitates the collection of comprehensive reservoir data. This data must include details concerning fracture features, porosity, permeability, and fluid properties. In addition, numerical simulation tools are frequently utilized in order to solve the governing equations and model the movement of fluid within the reservoir.

Due to the complexity of the model and the amount of computation it requires, specialized software as well as knowledge and experience in reservoir engineering and simulation techniques are required.

## Finite Element Model:

In pressure transient analysis, the application of the powerful numerical simulation approach known as the finite element model (FEM) is utilized to investigate reservoir dynamics. By partitioning the reservoir into numerous discrete parts, also known as cells, a more accurate picture of the storage space is produced. It is possible to obtain a more precise description of the reservoir's heterogeneity and anisotropy by assigning attributes such as permeability, porosity, and rock compressibility to each individual cell. This approach is taken. For the purpose of simulating fluid flow and pressure behavior across the entirety of the reservoir, the FEM takes into account the governing fluid flow equations, such as Darcy's law and the continuity equation, and then solves these equations numerically for each element.

In order to better understand how the finite element model functions in pressure transient analysis, the following step-by-step explanation has been provided:

> 1. Reservoir Discretization: The first step in building a finite element model is to discretize the reservoir into a mesh of interconnected elements or cells. The mesh can be structured (rectangular or triangular) or unstructured

(irregular-shaped elements). The choice of mesh type depends on the reservoir geometry and complexity.

2. Assigning Properties: Each element in the mesh is assigned properties such as permeability, porosity, and rock compressibility. These properties can vary from cell to cell to account for reservoir heterogeneity. The data for these properties are typically obtained from core samples, well logs, or other reservoir characterization techniques.

3. Boundary Conditions: The next step is to define boundary conditions that represent the reservoir's boundaries or interfaces. This includes specifying the pressure or flow rate at the boundaries or applying constraints to simulate no-flow boundaries. The boundary conditions reflect the reservoir's connectivity and interaction with external systems.

4. Fluid Flow Equations: The FEM solves the governing fluid flow equations for each element. These equations include Darcy's law, which describes fluid flow through porous media, and the continuity equation, which accounts for the conservation of mass. The fluid flow equations are modified to include any additional complexities, such as anisotropy, multiphase flow, or compositional effects.

## Darcy's Law

The movement of fluid through a porous media is characterized by Darcy's law, which is a fundamental principle in the field of fluid mechanics. Within the porous media, it offers a mathematical relationship between the flow rate, fluid velocity, and pressure gradient. The application of Darcy's rule is commonplace in a variety of domains, such as hydrogeology, petroleum engineering, and soil mechanics, among others.

The mathematical expression of Darcy's law is as follows:

$$Q = -kA(dP/dx)$$

Where:

- Q represents the volumetric flow rate of the fluid through the porous medium (measured in cubic meters per second or barrels per day).

- k denotes the permeability of the porous medium, which is a measure of how easily the fluid can flow through it (measured in Darcies or millidarcies).

- A represents the cross-sectional area perpendicular to the flow direction (measured in square meters or square feet).

- (dP/dx) represents the pressure gradient or pressure change per unit distance along the flow direction (measured in Pascal per meter or psi per foot).

Here is a detailed explanation of the components of Darcy's law:

1. Flow Rate (Q): The flow rate represents the volume of fluid that passes through a unit area per unit time. It is a measure of the quantity of fluid flowing through the porous medium. The negative sign in Darcy's law indicates that the flow is in the direction of

decreasing pressure.

2. Permeability (k): Permeability is a property of the porous medium that characterizes its ability to transmit fluids. It quantifies the ease with which the fluid can flow through the medium. A higher permeability indicates a more porous and permeable medium, allowing fluid to flow more easily. Permeability is determined by factors such as the size, shape, and connectivity of the pore spaces within the porous medium.

3. Cross-Sectional Area (A): The cross-sectional area refers to the area perpendicular to the direction of fluid flow. It represents the effective flow area available for fluid movement through the porous medium. The larger the cross-sectional area, the higher the flow rate for a given pressure gradient.

4. Pressure Gradient ((dP/dx)): The pressure gradient represents the rate of change of pressure per unit distance along the direction of flow. It indicates the pressure difference between two points divided by the distance between them. A higher pressure gradient corresponds to a steeper pressure drop per unit distance, resulting in increased fluid flow.

In order for Darcy's law to be valid, it is necessary to assume that the flow through the porous material is laminar and that the fluid adheres to Newton's law of viscosity. It is based on a number of assumptions that are intended to simplify things, such as the lack of turbulence, the effects of compressibility, and changes in the characteristics of the fluid.

The implementation of Darcy's law occurs frequently in the real world. It is helpful in estimating groundwater flow rates and determining the migration of contaminants within aquifers, for instance, in the field of hydrogeology. The application of Darcy's law in petroleum engineering allows for the analysis of fluid flow in reservoirs, the determination of reservoir parameters, and the optimization of well performance. In the field of soil mechanics, it is also helpful in understanding how water moves through the soil and determining how stable the soil is.

In general, Darcy's law is a fundamental mathematical connection that defines the flow of fluid through porous surfaces. This equation enables the quantitative study of flow rates and pressure gradients in a variety of different applications.

## The Bernoulli's Equation

In the field of fluid dynamics, the Bernoulli equation is a fundamental equation that connects the pressure, velocity, and elevation of a fluid in steady, incompressible flow. It stems from the idea that energy should be conserved along a streamline as much as possible. The equation of Bernoulli can be written as follows:

$$P + 1/2\rho v^2 + \rho gh = \text{constant}$$

Where:

- P represents the pressure of the fluid (measured in Pascal or psi).

- $\rho$ denotes the density of the fluid (measured in $kg/m^3$ or $lb/ft^3$).

- v represents the velocity of the fluid (measured in m/s or ft/s).

- g denotes the acceleration due to gravity (measured in $m/s^2$ or $ft/s^2$).

- h represents the elevation of the fluid above a reference point (measured in meters or feet).

According to the equation, the total energy of a fluid, which takes into account the pressure energy, kinetic energy, and potential energy, stays the same along a streamline even if there is no additional work done or energy lost from the environment. This includes all three types of energy. This equation is helpful when studying fluid flow in pipes, nozzles, and other flow devices, all of which have the potential for the fluid's pressure, velocity, and elevation to change.

## The Reynolds Number Equation:

The flow regime of a fluid, such as whether it is laminar or turbulent, can be determined with the help of an equation called the Reynolds number equation. It establishes a connection between the fluid flow's inertial forces and the viscous forces that act upon it. The Reynolds number, sometimes known as Re, can be defined as follows:

$$Re = (\rho v D) / \mu$$

Where:

- $\rho$ represents the density of the fluid (measured in $kg/m^3$ or $lb/ft^3$).

- v denotes the velocity of the fluid (measured in $m/s$ or $ft/s$).

- D represents a characteristic length or diameter associated with the flow (measured in meters or feet).

- $\mu$ denotes the dynamic viscosity of the fluid (measured in $Pa \cdot s$ or $lb/ft \cdot s$).

There is no way to express the Reynolds number in terms of its dimensions. It acts as an indicator of the behavior of the flow, with distinct ranges corresponding to the different types of flow (laminar, turbulent, etc.). Laminar flows are those that have a low Reynolds number (Re less than 2,000), and they are characterized by the movement of the fluid in smooth layers with very little mixing. Flows that have a high Reynolds number (Re > 4,000) are regarded to be turbulent. These flows are distinguished by their disorderly fluid motion and increased mixing. It is dependent on the particular system as well as the conditions to determine the transitional range between laminar and turbulent flow.

The Reynolds number equation is critical for anticipating flow regimes and finding the suitable correlations and equations to employ in fluid flow calculations. This is because the Reynolds number is a measure of the resistance to flow. Taking into account the flow characteristics as well as the accompanying pressure drops and losses, it provides assistance to engineers in the process of designing and analyzing flow systems such as pipelines, channels, and airfoils.

5. Numerical Solution: The finite element method discretized the governing equations for fluid flow and pressure behavior into algebraic equations. Various numerical techniques, such as the finite difference or finite volume methods, are employed to solve these equations iteratively. The solution is obtained by updating the pressure values within each element based on neighboring element properties and boundary conditions.

6. Pressure Transient Analysis: Once the numerical solution is obtained, the FEM generates pressure responses over time at different locations within the reservoir. These pressure responses can be compared to the observed pressure data from well tests to calibrate the

model and estimate reservoir properties. The comparison involves adjusting model parameters, such as permeability or porosity, until a satisfactory match is achieved between the modeled and observed pressure responses.

Worked Example:

Consider for a moment a fictitious reservoir that has intricate heterogeneity and erratic boundary conditions. In order to accurately represent the asymmetrical structure of the reservoir, it has been discretized into a finite element mesh comprised of triangle elements.

The data used to characterize the reservoir are used to determine how each element should be characterized in terms of its qualities, such as permeability and porosity. The boundary conditions have been defined, taking into account the existence of a barrier with no flow on one side and a border with constant pressure on the other side.

Using the finite element approach, the equations describing the flow of fluid, including Darcy's law and the continuity equation, are solved numerically for each element in the system. The solution will iteratively update the pressure values contained within each element, taking into consideration the attributes of neighboring elements and the boundary conditions.

The pressure responses at various sites within the reservoir are created throughout time once the numerical solution has been obtained. These pressure responses are examined in light of the data gathered from well tests on the measured pressure.
Imagine if the pressure response that was modeled did not initially match the data that was observed. In this scenario, the properties of the reservoir,

such as its permeability, porosity, or boundary conditions, are altered through a series of iterative processes until a match that is suitable is found. The goal of this iterative procedure is to reduce the discrepancy between the pressure responses that were modeled and those that were observed through the use of optimization techniques.

The final calibrated finite element model provides insights into the behavior of the reservoir. These insights include fluid flow patterns, pressure distribution, and the impact of heterogeneity and irregular boundaries on the performance of the reservoir.

With the help of this data, more educated choices may be made about reservoir management techniques, production optimization, and well placement.

By discretizing the reservoir into finite elements, assigning properties, numerically solving fluid flow equations, and comparing the modeled pressure responses with observed data, the finite element model (FEM) provides a detailed understanding of reservoir behavior, particularly in complex reservoirs with heterogeneity and irregular boundary conditions. In conclusion, the finite element model is a sophisticated numerical simulation technique used in pressure transient analysis to study reservoir behavior.

## Compositional Reservoir Model

The compositional reservoir model is a sophisticated method that is used to analyze the behavior of a reservoir when there are multiple fluid components present and their compositions change as fluids flow through the reservoir. This model is used to determine how a reservoir will behave when multiple fluid components are present. This model takes into consideration the phasing behavior as well as the compositional fluctuations of the fluids. It does this by taking into account elements including fluid characteristics, phase equilibrium, and fluid interactions. The compositional reservoir model is able to provide a more accurate representation of reservoir behavior because it takes into account the complexities of fluid composition. This is especially true in reservoirs that have fluid compositions that are particularly complex, such as gas-condensate reservoirs or reservoirs that are undergoing enhanced oil recovery processes.

To understand the working of a compositional reservoir model, let's consider an example:

Example: Gas-Condensate Reservoir Analysis

## 1. Fluid Characterization:

In a gas-condensate reservoir, the fluids typically consist of a mixture of hydrocarbon gasses (such as methane, ethane, propane, etc.) and liquid hydrocarbons (such as pentane, hexane, etc.). The compositional reservoir model requires accurate characterization of these fluids, including the composition, phase behavior, and fluid properties. This information is obtained through laboratory measurements and fluid analysis.

## 2. Equation of State:

The compositional reservoir model uses an equation of state (EOS) to describe the phase behavior of the fluids in the reservoir. An EOS is a mathematical model that relates pressure, temperature, and composition to calculate the properties of the fluids, including phase equilibrium and volumetric behavior. Commonly used EOS models include the Peng-Robinson, Soave-Redlich-Kwong, or the cubic equations of state. The EOS is essential for predicting fluid behavior and phase changes as the reservoir pressure and temperature conditions change.

## 3. Reservoir Simulation:

In the compositional reservoir model, the reservoir is discretized into a grid of cells, and fluid flow and transport equations are solved numerically for each cell. The model accounts for the flow of multiple fluid components, their interactions, and the phase changes that occur within the reservoir. The governing equations include mass conservation, energy conservation, phase equilibrium, and fluid flow equations. The numerical solution of these equations provides information about fluid saturation, phase distribution, and composition changes as fluids flow through the reservoir.

## 4. Fluid Flow and Phase Behavior:

The compositional reservoir model calculates the flow rates and pressure drops of the different fluid components within the reservoir. It accounts for phase behavior phenomena such as condensation, vaporization, and retrograde condensation, which occur in gas-condensate reservoirs. These phase behavior changes impact the fluid compositions, fluid saturations, and relative permeabilities within the reservoir.

5. Production Forecasting and Sensitivity Analysis:

Once the compositional reservoir model is calibrated and validated using available pressure and production data, it can be used for production forecasting and sensitivity analysis. The model predicts the behavior of different fluid components over time, allowing engineers to estimate production rates, fluid compositions, and recovery factors. Sensitivity analysis helps identify the key parameters and uncertainties that influence reservoir performance, such as the impact of changing operating conditions or fluid composition variations on production rates and ultimate recovery.

In order to effectively depict the behavior of many fluid components that are contained within a reservoir, the compositional reservoir model demands a substantial amount of data as well as computer resources. Throughout manufacturing, it offers extremely helpful insights into the behavior of fluids, the phase changes that occur, and the fluid compositions. Optimizing production plans, creating increased oil recovery techniques, and making educated decisions on reservoir development and management are all made easier with the assistance of the model.

Please take into consideration that the following illustration only offers a general summary of the compositional reservoir model. Expertise in reservoir engineering, fluid characterization, and numerical simulation techniques are required in order to carry out the implementation and analysis of a compositional reservoir model for a particular field or reservoir.

These intricate models not only provide more accurate representations of the behavior of reservoirs in the actual world, but they also make it possible to conduct an analysis of pressure transient data that is more in-depth. Nevertheless, in order to apply them, you will need significantly more data, computational resources, and knowledge than you would for more straightforward models such as the radial flow model. The specific properties of the reservoir, the goals of the research, and the data that are available all play a role in the decision of which model to use.

# Types of Well Tests

Well tests are an essential component in the process of determining the behavior and properties of subsurface reservoirs, which is an important part of the engineering discipline known as reservoir engineering. The results of these studies provide invaluable insight into the characteristics of the reservoir, including its permeability, porosity, and fluid flow rates. Buildup tests, drawdown tests, and interference tests are some of the most frequent kinds of well tests that can be carried out. Let's investigate each of these examinations in greater depth:

## 1. Buildup Tests:

Buildup tests, also known as pressure buildup tests or shut-in tests, are performed to assess the recovery of the well's internal pressure after it has been shut off. The output of fluids from the well is stopped so that an increase in pressure may be measured while conducting a buildup test. This causes the pressure within the reservoir to rise. Engineers are able to determine crucial reservoir properties, such as the permeability, skin factor (a measure of damage near the wellbore), and the presence of borders or obstacles that are affecting the reservoir, by monitoring the pressure response over time.

## 2. Drawdown Tests:

Drawdown tests, also known as pressure decline tests or flow tests, include the active production of fluids from a well while monitoring the subsequent pressure fall. Other names for drawdown tests include flow tests and pressure decline tests. The measurement of the productivity index of the well is the primary purpose of a drawdown test. This index provides a quantitative representation of the relationship that exists between the production rate and the pressure drawdown. Using this information, one may more accurately evaluate the performance of the well, ascertain whether or not there is damage around the wellbore, and estimate reservoir qualities such as permeability.

3. Interference Tests:

Monitoring the pressure fluctuations in various wells located within a reservoir is required in order to conduct interference tests, which are also known as pressure interference tests and pulse tests. In order to evaluate the connectivity and communication between wells, these tests are performed. During an interference test, fluid is either injected into or created from one well, and the pressure response in other neighboring wells is monitored to determine how it reacts to the injected or produced fluid. Engineers are able to determine several aspects regarding the reservoir by evaluating the pressure data. Some of these parameters include the transmissibility across wells, the presence of faults or obstacles that affect fluid movement, and the overall connectivity of the reservoir.

All of these different well tests lead to a better knowledge of the behavior of reservoirs, help with the characterisation of reservoirs, and contribute to the optimization of production plans. They supply crucial data that enables engineers to make educated judgments on the methods of well completion, stimulation, and reservoir management.

# Buildup Tests

In order to conduct buildup tests, the well is first sealed off, which means that no fluids are extracted from the wellbore. This enables the reservoir's internal pressure to gradually increase throughout the course of the test. This test is helpful in establishing the parameters of the reservoir as well as measuring the production of the well. The following is a list of the most important steps and equations that are involved in a buildup test:

### 1. Shut-in Phase:

In this phase, the well is closed off from production, and the pressure buildup within the reservoir is monitored over a specified duration. The rate of pressure increase during the shut-in phase depends on reservoir properties, such as permeability, and the presence of boundaries or barriers.

### 2. Pressure Transient Analysis:

During the buildup test, pressure data is collected at regular time intervals. This pressure data is then used to analyze the transient behavior of the reservoir and calculate important parameters. The pressure data can be plotted on a log-log plot to identify characteristic pressure responses, such as radial flow or boundary-dominated flow.

### 3. Pressure Derivative Analysis:

The pressure derivative is calculated to further analyze the pressure response during the buildup test. The pressure derivative helps identify the flow regime and provides insight into reservoir properties. It is typically calculated using the following equation:

$$D(p)/D(t) = (1 / \alpha) * (dP / dt) * (t / P)$$

where:

$D(p)/D(t)$ = Pressure derivative

$\alpha$ = Dimensionless constant (depends on the flow regime)

$dP / dt$ = Time derivative of pressure

$t$ = Time

$P$ = Pressure

4. Analysis Techniques:

There are various analysis techniques employed to interpret the pressure buildup test data and determine reservoir parameters. Some commonly used techniques include:

a. Horner Plot Analysis:

The Horner plot is a semi-log plot of pressure change versus logarithm of time. It helps estimate the average reservoir pressure, the skin factor, and the reservoir permeability. The Horner plot can be linearized for different flow regimes (radial, linear, or bilinear) to estimate the parameters.

b. Type Curve Matching:

Type curves are pre-determined pressure response curves that represent different reservoir flow regimes. By matching the observed pressure data with the type curves, reservoir parameters like permeability and skin factor can be estimated.

c. Pressure Buildup Analysis Models:

Various analytical and numerical models, such as the radial diffusion model, the infinite-acting radial flow model, or the dual-porosity model, can be utilized to simulate and analyze the pressure buildup response. These models consider the reservoir's geometry, fluid properties, and flow mechanisms to estimate reservoir parameters.

The interpretation of the accumulation test data provides insights into the reservoir features that are at play, such as permeability, skin factor, and boundaries that are affecting the well. It is absolutely necessary to have this knowledge in order to optimize production methods, well completion designs, and decisions about reservoir management.

# Drawdown Tests

Drawdown tests, also known as pressure decline tests or flow tests, include the active production of fluids from a well while monitoring the subsequent pressure fall. Other names for drawdown tests include flow tests and pressure decline tests. The measurement of the productivity index of the well is the primary purpose of a drawdown test. This index provides a quantitative representation of the relationship that exists between the production rate and the pressure drawdown. Using this information, one may more accurately evaluate the performance of the well, ascertain whether or not there is damage around the wellbore, and estimate reservoir qualities such as permeability.

$Q = (kA\Delta P) / (\mu L)$

Where:

Q is the fluid flow rate (measured in volume per unit time),

k is the permeability of the reservoir,

A is the cross-sectional area perpendicular to the flow direction,

$\Delta P$ is the pressure drop across the length of the reservoir,

$\mu$ is the fluid viscosity, and

L is the length of the reservoir.

When conducting a drawdown test, it is common practice to study the decrease in pressure using procedures that are specific to pressure transient analysis. The Horner plot is a method that is frequently utilized, and it contributes to the estimation of the well's productivity index (PI) in addition to other reservoir metrics. The plot of the dimensionless pressure derivative versus time that is known as the Horner plot is a semilogarithmic plot.

The dimensionless pressure derivative (pd) is calculated using the following equation:

$pd = (\Delta P / \Delta t) / (\Delta P0 / (t0\textasciicircum m))$

Where:

ΔP is the change in pressure at a specific time,

Δt is the corresponding change in time,

ΔP0 is the initial pressure change at a reference time t0,

t0 is the reference time, and

m is the slope of the straight line on the Horner plot.

The slope of the Horner plot can be used to estimate the well's productivity index (PI) using the following equation:

$$PI = 0.00708Q / (\Delta P0 \; \mu \; B \; ln(rw \; / \; re))$$

Where:

Q is the flow rate,

ΔP0 is the initial pressure drop at the reference time,

μ is the fluid viscosity,

B is the fluid formation volume factor,

rw is the wellbore radius, and

re is the reservoir radius.

Engineers are able to analyze the performance of the well, determine whether there is damage or skin around the wellbore, and estimate reservoir parameters such as permeability by examining the Horner plot and

95

calculating the productivity index (PI).

For the purpose of attaining a more in-depth comprehension of the behavior exhibited by the reservoir in the course of drawdown testing, it may be necessary to make use of additional analysis methods, such as type curve matching and pressure buildup analysis.

It is essential to keep in mind that the equations and analysis methods described earlier are merely simplified approximations of the full picture. Depending on the particular reservoir and test conditions, there may be modifications and extra factors to take into consideration.

## Interference Tests

The interference tests that a well performs are extremely helpful in gaining an understanding of the connectivity and communication that exists between wells in a reservoir. They assist reservoir engineers in gaining an understanding of how fluid moves and interacts between multiple wells, which is essential for improving production techniques and reservoir management. During an interference test, the pressure changes that occur as a result of the production or injection of fluid in one well are monitored in wells that are located nearby.

When conducting an interference test, the primary purpose is to ascertain the level of transmissibility between wells. The ease with which fluids can move across different compartments of a reservoir, such as between wells, is referred to as the transmissibility of the reservoir, and it is denoted by the sign "T." Both the permeability of the rock that makes up the reservoir and the qualities of the fluid can have an effect on it. Engineers are able to make an estimate of the transmissibility and gain a better understanding of the connectivity of the reservoir by evaluating pressure data from an interference test.

The principle of superposition is commonly used by engineers to analyze the pressure data that was obtained from an interference test. This principle states that the pressure response in a reservoir due to multiple sources or sinks can be obtained by adding the individual responses from each source or sink. Because of this approach, engineers are able to estimate the transmissibility between wells by basing their calculations on the pressure changes that were recorded in many wells at the same time as the interference test.

The Theis equation, which explains the pressure response in a homogeneous and isotropic reservoir owing to a constant-rate fluid source, is an equation that is frequently used for the analysis of interference tests. The Theis equation is based on the assumption of radial flow in a reservoir that is homogenous and does not contain any limits or barriers that could alter fluid flow. This is the equation that needs to be solved:

$$p(r, t) = p\_i - (Q / (4 * \pi * T)) * W(u),$$

where:

- $p(r, t)$ is the pressure at radial distance r from the source well at time t.

- $p\_i$ is the initial pressure in the reservoir before the interference test.

- Q is the constant-rate injection or production rate from the source well.

- T is the transmissibility between the source well and the observation well.

- $W(u)$ is the well function, which is a mathematical function that depends on the dimensionless parameter u.

The dimensionless parameter u is defined as:

$$u = (r^2 * S) / (4 * T * t),$$

where:

- r is the radial distance from the source well to the observation well.

- S is the strativity of the reservoir, which represents its ability to store fluids.

Engineers are able to come up with an estimate of the transmissibility T between wells by applying the Theis equation to the pressure data that was recorded during the interference test.

It is essential to keep in mind that real reservoirs are typically heterogeneous and anisotropic, which means that the rock characteristics and the behavior of fluid flow might vary in different directions. This fact is vital to keep in mind. In these kinds of circumstances, accurate analysis and interpretation of interference test data may call for the use of more involved mathematical models and numerical simulations. The impacts of reservoir heterogeneity, anisotropy, boundaries, and other factors that may influence fluid flow between wells can be incorporated into these more complex models.

The information that may be gleaned from interference testing regarding the connectivity and communication of wells in a reservoir is extremely valuable. Reservoir engineers are able to evaluate transmissibility, identify barriers or faults that alter fluid flow, and enhance reservoir management strategies for optimal production by analyzing pressure data using proper mathematical models.

# Flow Regimes

During the testing of a well, multiple flow regimes may occur based on the properties of the reservoir and the way fluids behave within it. These flow regimes are determined by the features of the reservoir. Radial flow, linear flow, and bilinear flow are the three primary types of flow regimes that can be experienced. The behavior of the pressure is altered in a manner that is distinctively influenced by each of these flow regimes. Let's go through each of them one at a time:

1. Radial Flow:

Radial flow occurs when fluid flows radially from the reservoir towards the wellbore. In this regime, the reservoir behaves as if it is infinite and homogeneous, and the pressure distribution around the wellbore is symmetric. Initially, during the early time period, radial flow dominates the pressure behavior. The pressure declines rapidly as the wellbore storage is depleted. The pressure behavior in this regime is typically characterized by a slope of -1/2 on a log-log plot of pressure versus time.

2. Linear Flow:

Linear flow occurs when the pressure transient is influenced by the limited conductivity of the reservoir near the wellbore. It usually happens in reservoirs with low permeability or when the well is in contact with a low-permeability formation, such as fractures or tight rock. In linear flow, the pressure distribution around the wellbore is no longer symmetric, and the pressure decline becomes more gradual compared to radial flow. The pressure behavior in this regime is characterized by a slope of -1 on a log-log plot of pressure versus time.

3. Bilinear Flow:

Bilinear flow occurs when a well test exhibits a transition from radial flow to linear flow behavior. This transition happens when the pressure transient reaches a certain radial distance in the reservoir, beyond which linear flow dominates. The pressure behavior during bilinear flow is a combination of the behaviors observed in radial and linear flow. Initially, the pressure decline follows the -1/2 slope characteristic of radial flow, and later transitions to a -1 slope characteristic of linear flow. The transition point between the two regimes is known as the bilinear time.

Several flow regimes may be seen during the testing of a well because of the characteristics of the reservoir and the way fluids behave when they are contained inside it. The characteristics of the reservoir are what decide the flow regimes that are there. The three basic flow regimes that a person may encounter are radial flow, linear flow, and bilinear flow. Radial flow is the most common form. Each of these flow regimes has a different effect on the way the behavior of the pressure is adjusted, and each of these flow regimes has their own unique influence. Let's go through each one of them in turn, starting with the easiest one:

# Radial Flow

Let's delve into radial flow in further detail.

Fluid moves in a radial direction from the reservoir toward the wellbore when radial flow is used, and the pressure distribution remains symmetrical around the wellbore. Radial flow is one of two types of flow. This flow regime operates under the presumption that the reservoir is both infinite and homogenous. What this means is that the reservoir's characteristics, such as permeability and porosity, are assumed to be the same all the way through the reservoir.

During the early time period of radial flow, the behavior of the pressure is dominated by the depletion of wellbore storage. This continues until the later time periods of radial flow. When the well is first opened, the pressure inside the wellbore and in the area immediately surrounding the wellbore is higher than the typical pressure in the reservoir. The pressure drops precipitously as more and more of the wellbore storage is used up. This part of the process is referred to as the "wellbore storage" or "wellbore storage effect" phase.

The radial diffusion equation, which is also referred to as the radial flow equation, is the mathematical model that is utilized the vast majority of the time when analyzing radial flow. This equation, which estimates crucial reservoir parameters when solved, represents the behavior of the pressure during radial flow and can be used to describe that behavior.

The radial diffusion equation is given by:

$$\partial p/\partial t = (1/r)\, \partial/\partial r(r\, \partial p/\partial r) + S/4\mu kt,$$

where:

- p is the pressure in the reservoir (psi or Pa),

- t is time (hours or seconds),

- r is the radial distance from the wellbore (ft or meters),

- S is the storativity of the reservoir (dimensionless),

- $\mu$ is the viscosity of the fluid (cp or Pa·s),

- k is the permeability of the reservoir (md or m²),

- and t is the transmissibility of the reservoir (md-ft or m²-Pa).

The radial flow of fluid through the reservoir can be described with the help of this partial differential equation. The radial flow component is represented by the first term on the right-hand side of the equation, while the storage component is represented by the second term.

By solving the radial diffusion equation, we are able to determine the behavior of the pressure and deduce crucial parameters about the reservoir. While performing an analysis of well testing, the pressure response is frequently depicted on a log-log graph, with pressure being represented on the y-axis and time being represented on the x-axis. When radial flow predominates in the early time period, the pressure drop corresponds to a straight line on the log-log plot with a slope of -1/2. This indicates that the flow is counterclockwise.

The fact that the slope is -1/2 suggests that the rate of pressure drop is proportional to the square root of time shows that the slope. This behavior, which is characteristic of radial flow, is referred to as "linear flow" or "square root of time" behavior. Both of these terms refer to the same thing. Reservoir engineers are able to make estimates of factors such as permeability, skin factor, and wellbore storage by doing an analysis of the slope and intercept of the straight line.

It is important to keep in mind that radial flow assumptions are correct only in

situations in which the distance between the wellbore and the borders of the reservoir is sufficiently high in comparison to the size of the reservoir. As the well test progresses, the flow regime may change to either linear flow or bilinear flow, depending on the properties of the reservoir and the way fluids behave inside it.

These changes are determined by the characteristics of the flow.

In the following section, we are going to look into linear flow and bilinear flow in greater depth.

# Linear Flow

To understand the pressure behavior during linear flow, we need to consider Darcy's law, which describes the flow of fluid through porous media. Darcy's law is given as:

$$Q = -kA(dP/dx) / \mu$$

where:

Q is the flow rate of fluid,

k is the permeability of the reservoir,

A is the cross-sectional area perpendicular to the flow direction,

(dP/dx) is the pressure gradient along the flow direction,
103

and μ is the viscosity of the fluid.

In linear flow, the pressure fall is more gradual than it is in radial flow because the conductivity of the reservoir is lower near the wellbore in linear flow than it is in radial flow. As a direct consequence of this, the pressure gradient, denoted by dP/dx, is no longer held at a constant value but instead changes in accordance with the radial distance from the wellbore.

To analyze linear flow, we make use of the dimensionless pressure derivative, which provides valuable insights into the flow regime. The dimensionless pressure derivative is defined as:

$$qD = - (1 / P) (dP / dt)$$

where:

qD is the dimensionless pressure derivative,

P is the pressure, and

t is time.

On a log-log plot showing qD in relation to time, the dimensionless pressure derivative has a slope of -1 when linear flow is being considered. The behavior of linear flow is represented by this slope of -1, which demonstrates the characteristic behavior of linear flow. When the pressure transient reaches a particular radial distance in the reservoir, beyond which linear flow prevails, the linear flow regime often becomes apparent at later times. This is the case because linear flow dominates when the pressure transient reaches this distance.

The fact that the slope is negative suggests that the decrease in pressure is inversely proportional to the square root of the passage of time. It is possible to formulate it in mathematical terms as:

$$dP / dt \propto -1 / \sqrt{t}$$

Integrating this relationship gives:

$$P = P\_i - (C \sqrt{t})$$

where:

P is the pressure at a given time,

P_i is the initial pressure,

C is a constant related to the reservoir and fluid properties,

and t is the time.

When there is linear flow, the behavior of the pressure is no longer symmetric around the wellbore in the same way that it is when there is radial flow. Because of the reduced conductivity near the wellbore, the pressure distribution becomes more elongated along the flow direction. This indicates that the influence of this phenomenon can be seen.

Reservoir engineers are able to estimate significant parameters like permeability, skin factor (which measures the near-wellbore damage or enhancement), and reservoir boundaries by conducting an analysis of the pressure behavior during linear flow. Understanding linear flow is also helpful in optimizing production methods and making informed decisions on well performance and reservoir management in low-permeability reservoirs or formations with limited conductivity. This is because linear flow is a one-dimensional process.

## Bilinear Flow

As the pressure behavior changes from radial flow to linear flow during well testing, a crucial regime known as bilinear flow can be detected. Bilinear flow is characterized by the presence of both radial and linear flow. This change takes place when the pressure transient in the reservoir reaches a particular radial distance known as the bilinear radius or the radius of investigation. Both of these terms refer to the same radial distance. After reaching this distance, the pressure behavior will be dominated by linear flow since the limited conductivity of the reservoir near the wellbore will have disappeared.

Let's go into the mathematical features and equations that regulate this flow regime so that we can have a deeper comprehension of bilinear flow and its many nuances:

1. Radial Flow:

   During the initial stage of a well test, the pressure behavior is primarily governed by radial flow. Radial flow occurs when fluid flows radially from the reservoir towards the wellbore, assuming the reservoir is infinite and homogeneous. In this regime, the pressure decline is relatively rapid due to the large reservoir volume contributing to flow.

The pressure behavior during radial flow can be described by the radial diffusivity equation, also known as the diffusivity equation in cylindrical coordinates:

(1) $\Delta p / \Delta t = (1/r) * (\Delta / \Delta r) (r * k * \Delta p / \Delta r)$

where:

$\Delta p / \Delta t$ is the pressure change with respect to time,

r is the radial distance from the wellbore,

k is the permeability of the reservoir,

$\Delta p / \Delta r$ is the pressure gradient.

By solving the radial diffusivity equation, the pressure behavior during radial flow can be expressed as:

(2) $p(r, t) = p\_i - (0.5 * q\_s * \mu * c\_t * \ln(t/t\_i)) / (4 * \pi * k * h)$

where:

$p(r, t)$ is the pressure at radial distance r and time t,

$p\_i$ is the initial reservoir pressure,

$q\_s$ is the constant production rate,

$\mu$ is the fluid viscosity,

$c\_t$ is the total compressibility,

107

t_i is the initial shut-in time,

h is the reservoir thickness.

In the early time period, the pressure decline during radial flow follows a characteristic slope of -1/2 on a log-log plot of pressure versus time.

## 2. Linear Flow:

Linear flow occurs when the pressure behavior is influenced by the limited conductivity near the wellbore. This regime is typically observed in reservoirs with low permeability or when the well is in contact with a low-permeability formation, such as fractures or tight rock.

The pressure behavior during linear flow can be described by the linear flow equation:

$$(3) \; \Delta p / \Delta t = (k * h * q\_s) / (\mu * r * \ln(re/r))$$

where:

re is the external boundary radius.

Solving the linear flow equation provides the following expression for pressure behavior:

(4) $p(r, t) = p\_i - (q\_s * \mu * c\_t * h * \ln(t/t\_i)) / (2 * \pi * k * r)$

In the linear flow regime, the pressure decline follows a characteristic slope of -1 on a log-log plot of pressure versus time.

3. Bilinear Flow:

Bilinear flow occurs when the pressure transient transitions from radial flow to linear flow. The transition point, known as the bilinear time (t_bl), marks the moment when the linear flow dominates beyond a certain radial distance in the reservoir.

The pressure behavior that is noticed during bilinear flow is a combination of the characteristics that are seen during radial flow and linear flow. At first, the pressure decrease conforms to the -1/2 slope that is typical of radial flow; however, as time passes, it changes to the -1 slope that is typical of linear flow.

It is common practice to employ a plot of pressure derivative (dp/dt) versus time when doing an analysis of the bilinear flow regime. The pressure derivative plot displays different patterns that are helpful in identifying the transition from radial flow to linear flow and in predicting reservoir parameters. These patterns can be seen as a representation of the pressure as a function of time.

It is essential to keep in mind that the mathematical equations that have been presented up to this point are only simplified approximations of the actual flow behavior that occurs in a reservoir. Reservoirs that are found in the real world may display a behavior that is more complicated, such as the presence of barriers, reservoir heterogeneity, anisotropy, or other flow mechanisms. In spite of this, having a firm grasp on the core concepts of radial flow, linear flow, and bilinear flow is helpful in assessing the performance of the reservoir and interpreting the results of well tests.

# Wellbore Storage and Skin Effect

Wellbore storage and skin effect are two important concepts in well testing analysis that influence the behavior of pressure near the wellbore. Let's explore each concept in detail:

## Wellbore Storage

During a well test, the term "wellbore storage" is used to refer to the volume of fluids that are held in the wellbore itself. When a well is closed or shut-in for a period of time before the start of a well test, the pressure in the wellbore and near the wellbore begins to build up due to the accumulation of fluids. The fact that fluids can be compressed and that the wellbore has a restricted flow capacity both contribute to the development of this pressure buildup.

The initial phases of a well test are characterized by the depletion of wellbore storage, which in turn has an effect on the pressure response. When a well is opened for production or pressure monitoring, the initially high pressure in the wellbore begins to decrease. This is because the well is gaining access to more space. This decrease happens as a result of fluids that have been held in the wellbore beginning to flow into the reservoir. The wellbore volume, the formation's compressibility, and the fluid characteristics all play a role in determining how quickly wellbore storage is depleted. Other aspects include the fluid properties.

In order to analyze the effects of wellbore storage on pressure behavior, an additional term known as the wellbore storage constant (Cw) is incorporated into the pressure transient analysis equations. This term is used to account for the impact that wellbore storage has on pressure behavior. The wellbore storage constant is a representation of the relationship that exists between the change in pressure and the change in the cumulative production or pressure measurement:

$$\Delta p = C_w * \Delta q$$

111

where:

$\Delta p$ is the change in pressure,

$\Delta q$ is the change in cumulative production or pressure measurement.

While attempting to correctly interpret well test data and arrive at an accurate estimate of reservoir properties, it is essential to take into account wellbore storage. Ignoring the storage capacity of the wellbore might lead to inaccurate estimates and interpretations of the data.

During a well test, let's go through a detailed example of how to calculate the amount of storage available in the wellbore.

Consider the following scenario: we are conducting a well test in which we first stop production at the well for a set amount of time before restarting it. The measurements of pressure are collected at predetermined intervals of time, and we wish to estimate the wellbore storage constant (Cw) so that we can take into account the fluids that have been stored in the wellbore.

Here are the given data for the well test:

Initial shut-in time ($t\_i$): 6 hours

Initial pressure ($p\_i$): 5000 psi

Production time (t): 24 hours

Pressure measurements:

- t = 0 hours: p = 5000 psi

- t = 6 hours: p = 4800 psi

112

- t = 12 hours: p = 4600 psi

- t = 24 hours: p = 4200 psi

To calculate the wellbore storage constant (Cw), we need to determine the change in pressure ($\Delta p$) and the change in cumulative production ($\Delta q$) during the well test.

1. Calculate the change in pressure ($\Delta p$):

$$\Delta p = p - p\_i$$

- At t = 6 hours: $\Delta p\_6h$ = 4800 psi - 5000 psi = -200 psi

- At t = 12 hours: $\Delta p\_12h$ = 4600 psi - 5000 psi = -400 psi

- At t = 24 hours: $\Delta p\_24h$ = 4200 psi - 5000 psi = -800 psi

2. Calculate the change in cumulative production ($\Delta q$):

Since we have pressure measurements at specific time intervals, we can estimate the change in cumulative production by assuming a constant production rate (q_s) during each time interval.

- From t = 0 hours to t = 6 hours: q_s1 = (p_i - p_6h) / (Cw * t_i)

  $\Delta q1 = q\_s1 * t\_i$

113

- From t = 6 hours to t = 12 hours: q_s2 = (p_6h - p_12h) / (Cw * (t_i + 6 hours))

$\Delta$q2 = q_s2 * (t_i + 6 hours)


- From t = 12 hours to t = 24 hours: q_s3 = (p_12h - p_24h) / (Cw * (t_i + 12 hours))

$\Delta$q3 = q_s3 * (t_i + 12 hours)


Note: The wellbore storage constant (Cw) is the same for all time intervals.


3. Solve for the wellbore storage constant (Cw):


Using the estimated changes in pressure ($\Delta$p) and cumulative production ($\Delta$q) for each time interval, we can solve for the wellbore storage constant (Cw).


$\Delta$p = Cw * $\Delta$q


- For the first time interval: $\Delta$p_6h = Cw * $\Delta$q1

Cw = $\Delta$p_6h / $\Delta$q1


- For the second time interval: $\Delta$p_12h = Cw * $\Delta$q2

Cw = $\Delta$p_12h / $\Delta$q2

114

- For the third time interval: $\Delta p\_24h = Cw * \Delta q3$

$Cw = \Delta p\_24h / \Delta q3$

Average the calculated values of Cw from each time interval to obtain the final wellbore storage constant estimate.

Let's plug in the given data and perform the calculations:

1. Calculate the change in pressure ($\Delta p$):

$\Delta p\_6h = -200$ psi

$\Delta p\_12h = -400$ psi

$\Delta p\_24h = -800$ psi

2. Calculate the change in cumulative production ($\Delta q$):

$q\_s1 = (5000$ psi $- 4800$ psi$) / (Cw * 6$ hours$)$   [Assume constant production rate for the first interval]

$\Delta q1 = q\_s1 * 6$ hours

$q\_s2 = (4800$ psi $- 4600$ psi$) / (Cw * (6$ hours $+ 6$ hours$))$   [Assume constant production rate for the second interval]

$\Delta q2 = q\_s2 * (6$ hours $+ 6$ hours$)$

115

q_s3 = (4600 psi - 4200 psi) / (Cw * (6 hours + 12 hours))   [Assume constant production rate for the third interval]

$\Delta$q3 = q_s3 * (6 hours + 12 hours)

3. Solve for the wellbore storage constant (Cw):

Cw = $\Delta$p_6h / $\Delta$q1

Cw = -200 psi / (q_s1 * 6 hours)

Cw = $\Delta$p_12h / $\Delta$q2

Cw = -400 psi / (q_s2 * (6 hours + 6 hours))

Cw = $\Delta$p_24h / $\Delta$q3

Cw = -800 psi / (q_s3 * (6 hours + 12 hours))

Average the values of Cw obtained from each time interval to get the final estimate.

By performing the above calculations, we can determine the wellbore storage constant (Cw) and account for the stored fluids in the wellbore during the well test analysis.

# Skin Effect:

The term "skin effect" refers to the modification of circumstances near the wellbore that is brought about by the processes of drilling, completion, and production. This demonstrates a departure from the behavior of fluid flow that would be appropriate for the wellbore. The skin effect may be either positive or negative, indicating either injury or stimulation in the area close to the wellbore, respectively.

A positive skin shows an increase in flow resistance or a decrease in permeability near the wellbore. This is denoted by a Skin value greater than 0. This may be the result of a number of different circumstances, including drilling mud invasion, formation damage, or the presence of compounds that cause skin irritation. A positive skin decreases the effective permeability of the formation, which in turn causes a decrease in the amount of fluid that enters the wellbore. As a direct consequence of this, the pressure in the vicinity of the wellbore is significantly higher than it would be in a formation that had not been damaged.

Negative skin, on the other hand, denotes a stimulating effect close to the wellbore, which leads to increased permeability or decreased flow resistance. This is denoted by the expression "Skin 0." This may take place as a result of procedures such as acidizing the wellbore, hydraulic fracturing, or wellbore cleansing. As a result of the increased inflow of fluids into the wellbore that negative skin facilitates, the pressure in the vicinity of the wellbore is lower than it would be in a formation that had not been damaged.

The skin effect is taken into account in the analysis of well testing thanks to the incorporation of a skin factor (S). The skin factor is incorporated in the pressure transient analysis equations and represents the additional pressure drop or pressure increase induced by the alteration of circumstances around the wellbore.

$$\Delta p = 2\pi kh * (q/(4\pi kt)) * \ln(re/r) + S * (q/(2\pi k)) * (1/\sqrt{(4\pi kt)})$$

where:

$\Delta p$ is the pressure drop,

k is the permeability,

h is the thickness of the reservoir,

q is the flow rate,

r is the radial distance from the wellbore,

re is the external boundary radius,

t is the time.

Let's walk through an in-depth example of calculating the skin effect in a well testing analysis.

Assumptions:

We have pressure and flow rate data from a well test.

The well is in a homogeneous reservoir.

We will use the pressure derivative plot to estimate the skin effect.

Step 1: Data Preparation

Gather the pressure and flow rate data from the well test.

Ensure that the pressure data is in a suitable format for analysis, such as pressure versus time or pressure versus cumulative production.

Step 2: Plot Pressure Derivative

Calculate the pressure derivative (dp/dt) from the pressure data. This can be done numerically by taking the difference between adjacent pressure data points and dividing by the corresponding time interval.

Plot the pressure derivative as a function of time.

Step 3: Identify the Linear Flow Period

Identify the linear flow period on the pressure derivative plot. This is characterized by a straight line with a slope of -1.

The linear flow period typically occurs after the initial radial flow period.

Step 4: Determine the Intercept

Determine the intercept of the linear flow period on the pressure derivative plot.

The intercept represents the value of the pressure derivative at a reference time.

Step 5: Calculate Skin Factor

The skin factor (S) can be estimated using the following equation:

$$S = (2.3 * Q * \mu * h) / (slope * k * A)$$

where:

Q is the flow rate,

$\mu$ is the fluid viscosity,

h is the reservoir thickness,

slope is the slope of the linear flow period on the pressure derivative plot,

k is the permeability,

A is a dimensionless constant related to the wellbore storage and typically ranges from 0.5 to 1.0.

Plug in the known values into the equation and calculate the skin factor.

Step 6: Interpret the Skin Effect

A skin factor that is positive suggests that there is damage or a reduction in permeability near the wellbore, whereas a skin factor that is negative shows that there is stimulation or an improvement in permeability near the wellbore.

Engineers are able to determine the level of near-wellbore damage or stimulation, estimate the skin factor value, and make educated judgments on wellbore treatment or production optimization by examining the pressure behavior and measuring the skin effect.

For an appropriate analysis of well testing, it is essential to have a solid understanding of both the wellbore storage and the skin effect, as these two factors play key roles in the interpretation of pressure behavior and the evaluation of reservoir performance near the wellbore.

# Pressure Derivative Analysis

The pressure derivative is a fundamental mathematical tool in well test analysis that is used to interpret pressure transient data collected from reservoirs during well testing.

This data can be obtained from the well. It offers helpful insights on the flow behavior of fluids within the reservoir, which makes it possible to estimate important parameters of the reservoir. The pressure derivative is obtained from the pressure-time data and is then used to establish the limits of the reservoir, identify the various flow regimes, and estimate parameters such as permeability and skin factor.

Let's go back to the fundamentals in order to have a grasp on the idea of pressure derivative. A fluid, often oil, gas, or water, is produced or injected into a well during a well test, and the pressure response at the wellbore is monitored over the course of the test. After collecting and analyzing this pressure transient data, the features of the reservoir can then be inferred.

Calculating the pressure derivative requires differentiating the data on the pressure with regard to the passage of time. The derivative of pressure, often known as dp/dt, is a mathematical term that refers to the rate at which pressure (p) varies with respect to the passage of time (t). While doing an analysis of a well test, the pressure derivative is frequently abbreviated as pD or simply D. It is a quantity that cannot be measured and is generated by normalizing the pressure derivative using appropriate variables in order to remove the influence of the well's qualities and the fluid's properties.

The pressure derivative offers extremely helpful insight into the flow pattern that is being experienced in the reservoir. Several flow regimes can be determined by studying the form and size of the pressure derivative

122

curve. This allows for the identification of distinct flow patterns. Radial flow, linear flow, bilinear flow, and pseudo-steady state flow are the types of flow regimes that fall under this category.

Understanding the many flow regimes that might occur inside a reservoir is essential to producing accurate characterizations of the reservoir. Each flow regime denotes a distinct pattern of fluid movement that can occur within the reservoir.

Radial flow happens in the beginning phases of a well test when the pressure response is dominated by flow from the reservoir into the wellbore. This is when radial flow occurs. On a log-log plot, the pressure derivative displays a characteristic slope of -1/2, which is characteristic of this regime. The radial flow regime is typically seen in reservoirs that have a homogeneous and isotropic composition.

When the pressure transient data display a straight line on a log-log plot, linear flow has taken place in the system. This flow regime often takes place in rocks that have an exceptionally low permeability or in wells that produce from fractured reservoirs.

When considering linear flow, the slope of the pressure derivative is -1. When both radial and linear flow are present at the same time, a bilinear pattern of flow can be seen. On a log-log plot, the slope of the pressure derivative curve abruptly shifts from -1/2 to -1, indicating a significant change in the relationship between the two variables. The presence of bilinear flow in a reservoir is frequently suggestive of the presence of borders or of reservoirs that are influenced by the wellbore storage effects.

When the reservoir pressure has reached a quasi-steady state as a result of depletion, the flow rate will change to a pseudo-steady state. During this phase of the experiment, the pressure derivative is very close to zero. When well tests are performed on reservoirs that have a high permeability or when there is a substantial distance between the well and the reservoir borders, it is typical to detect flows that behave as though they are in a pseudo-steady state.

The well test analysts are able to estimate the properties of the reservoir by first studying the pressure derivative curve, and then identifying the flow regimes. For instance, the permeability of the reservoir can be related to the slope of the pressure derivative in the radial flow regime, which is -1/2. The bilinear flow regime has the potential to shed light on reservoir boundaries as well as the existence of impacts that are close to the wellbore. It is possible to determine the skin factor in the pseudo-steady-state flow regime by making use of the pressure derivative. The skin factor depicts the augmentation or damage that occurs close to the wellbore.

For the purpose of identifying flow regimes, estimating reservoir attributes, and characterizing reservoir behavior, well test analysis makes use of a strong mathematical tool known as the pressure derivative. It supports the process of making educated decisions on the development of the reservoir as well as production plans by providing essential insights into the flow behavior of fluids contained within the reservoir.

Let's have a look at a straightforward example in order to clarify the idea of pressure derivatives. Let's say we have a cylindrical container that's been filled with a liquid, and we want to investigate how the pressure inside the container shifts in relation to the passage of time as the volume of the liquid is altered by either increasing or decreasing.

To get started, let's refer to the volume of the liquid as V and the pressure inside the container as P. This will help us visualize the relationship between the two. We are interested in determining how the pressure P varies in response to changes in the volume V. If we want to express this

124

relationship using mathematics, we can say that it is:

P = f(V)

where f is some function that relates the pressure to the volume.

Now, let's assume that we increase the volume of the liquid by a small amount $\Delta V$ and observe the corresponding change in pressure, $\Delta P$. This change in pressure can be written as:

$\Delta P = f(V + \Delta V) - f(V)$

The pressure derivative, denoted as dp/dv or dP/dV, represents the rate of change of pressure with respect to volume. It tells us how much the pressure changes for a given change in volume.

To calculate the pressure derivative, we need to consider the limit as $\Delta V$ approaches zero. In other words, we want to find the derivative of the function f(V) with respect to V. We can express this as:

$dP/dV = \lim(\Delta V \to 0) [\Delta P/\Delta V]$

Now, let's work through an example to illustrate this concept. Suppose we have a container filled with water, and we want to determine the pressure derivative at a specific volume. We measure the pressure at two different

125

volumes: V1 = 100 mL and V2 = 150 mL. The corresponding pressures are P1 = 200 kPa and P2 = 250 kPa, respectively.

Using these values, we can calculate the pressure derivative as follows:

$\Delta V$ = V2 - V1 = 150 mL - 100 mL = 50 mL = 0.05 L

$\Delta P$ = P2 - P1 = 250 kPa - 200 kPa = 50 kPa

Now, we can use these values in the formula for the pressure derivative:

dP/dV = lim($\Delta V$->0) [$\Delta P/\Delta V$]

In this particular illustration, we can see that V is not getting closer to zero; nonetheless, we are still able to make a guess. Let's take it for granted that V is sufficiently negligible that we can think of it as getting close to zero. In this scenario, we are able to determine the pressure derivative.

dP/dV = $\Delta P/\Delta V$ = 50 kPa / 0.05 L = 1000 kPa/L

Therefore, the pressure derivative in this scenario is 1000 kPa per liter (kPa/L). This value tells us that for every increase of 1 liter in volume, the pressure inside the container increases by 1000 kPa.

When working with complicated systems, it is essential to keep in mind that, in reality, the pressure derivative is typically computed by employing methods that are more accurate, such as numerical differentiation or regression analysis. The idea of pressure derivative can be made clearer with the assistance of this simple graphic, which uses this example.

# Numerical Simulation

In the oil and gas business, numerical simulation models are utilized extensively in order to mimic well testing and acquire a better knowledge of reservoir behavior. These models use mathematical methods to depict the complicated physical processes that are taking place in the subsurface. As a result, they provide useful insights into a variety of elements of reservoir performance.

When it comes to testing wells, numerical simulation models have numerous benefits that standard analytical models do not. They make it possible to take into account non-ideal effects, like the heterogeneity of the reservoir and the displacement of the fluid, which are frequently seen in reservoirs that are found in the actual world. The following is a list of important uses and benefits that can be gained from employing numerical simulation models for well testing:

1. Reservoir Heterogeneity: Reservoirs are rarely homogeneous, and their properties vary spatially. Numerical simulation models can incorporate detailed reservoir characterization, including variations in permeability, porosity, and fluid saturations across the reservoir. By simulating well tests in such heterogeneous reservoirs, engineers can observe the impact of these variations on pressure responses and flow behavior. This helps in assessing the reservoir's connectivity, identifying preferential flow paths, and optimizing production strategies.

2. Fluid Displacement and Multiphase Flow: Well tests often involve the production or injection of multiple fluids (e.g., oil, water, gas) simultaneously. Numerical simulation models can handle multiphase flow and accurately predict the behavior of different fluids during well testing. They can capture complex phenomena such as capillary pressure, relative permeability, and fluid

displacement mechanisms. This allows engineers to evaluate the efficiency of different production/injection strategies, understand the movement of fluids within the reservoir, and estimate the ultimate recovery of hydrocarbons.

3. Sensitivity Analysis and Optimization: Numerical simulation models enable engineers to conduct sensitivity analyses to assess the impact of various parameters on well test responses. By systematically varying parameters such as reservoir properties, well configurations, and fluid properties, they can evaluate how these factors influence the test results. Additionally, simulation models can be used to optimize well test design, including the determination of appropriate test durations, flow rates, and pressure measurement locations. This helps in reducing uncertainty and improving the accuracy of reservoir characterization.

4. Field-Scale Behavior: Unlike analytical models, numerical simulation models can simulate well tests at the field scale, considering the entire reservoir and all its wells. This capability allows for a comprehensive assessment of reservoir behavior, including the interactions between multiple wells and the effects of reservoir boundaries and geological features. By simulating large-scale well tests, engineers can evaluate the impact of reservoir architecture, boundaries, and other factors on production/injection rates, pressure responses, and ultimate recovery.

5. Future Performance Prediction: Numerical simulation models provide a forward-looking view of reservoir performance beyond the well test itself. By calibrating the model to match historical well test data, engineers can then use the validated model to forecast future production behavior, estimate the long-term productivity of the reservoir, and plan optimal field development strategies.

In summary, the application of numerical simulation models in simulating well tests offers a more detailed understanding of reservoir behavior, accounting for non-ideal effects such as reservoir heterogeneity and fluid displacement. These models allow engineers to optimize well test design, evaluate the impact of various factors, and forecast reservoir performance, leading to more informed decision-making in the oil and gas industry.

# Reservoir Heterogeneity

The term "reservoir heterogeneity" refers to the spatial changes in attributes that can be found within a reservoir. Some examples of these variations include permeability, porosity, and fluid saturations. These differences have a substantial impact on the behavior of fluid flow and can have a considerable influence on the results of well tests as well as the performance of the reservoir. When it comes to understanding and measuring the effects of reservoir heterogeneity on well testing and production plans, the role that numerical simulation models play is absolutely essential. A more in-depth explanation of reservoir heterogeneity and its consequences for well testing can be found as follows:

1. Definition and Types of Reservoir Heterogeneity:

Reservoir heterogeneity refers to the non-uniform distribution of reservoir properties across space. It arises due to various geological factors such as depositional environments, diagenesis, and structural features. Heterogeneity can manifest in different forms, including vertical and lateral variations in permeability, variations in porosity, and variations in fluid saturations.

- Vertical heterogeneity: This type of heterogeneity refers to variations in reservoir properties along the vertical direction, such as changes in permeability and porosity with depth. It can occur due to different sedimentary layers or stratigraphic features within the reservoir.

- Lateral heterogeneity: Lateral heterogeneity describes variations in reservoir properties in the horizontal plane. It can occur due to variations in sedimentary facies, depositional environments, or structural features like faults and fractures.

2. Incorporating Reservoir Heterogeneity in Numerical Simulation Models:

Numerical simulation models utilize a grid-based representation of the reservoir, dividing it into cells or blocks. To account for reservoir heterogeneity, these models assign different properties (permeability, porosity, fluid saturations) to each cell based on the geological characterization of the reservoir. This characterization is obtained from well logs, core samples, seismic data, and other available information.

The reservoir properties are often represented as spatially varying fields, where interpolation techniques are used to assign values to the grid cells. Different interpolation methods can be employed, such as simple averaging, kriging, or geostatistical techniques, to capture the spatial correlation of the reservoir properties.

3. Impact of Reservoir Heterogeneity on Well Tests:

When well tests are conducted in heterogeneous reservoirs, the variations in permeability, porosity, and fluid saturations can significantly influence pressure responses and fluid flow behavior. Some key aspects to consider are:

a. Pressure Transient Analysis: Reservoir heterogeneity affects the pressure behavior observed during well tests. Variations in permeability lead to contrasting flow rates in different parts of the reservoir, resulting in pressure variations. Analyzing pressure transient data using numerical simulation models allows engineers to identify preferential flow paths, areas of high permeability contrast, and reservoir connectivity.

b. Flow Behavior: Heterogeneity influences the movement and distribution of fluids within the reservoir during well testing. Channels of high permeability may dominate the flow, while low-permeability regions act as barriers or baffles. Numerical simulation models capture these flow patterns and provide insights into fluid displacement, breakthrough times, and sweep efficiency. This information helps optimize well placement, design appropriate production strategies, and mitigate unwanted effects such as early water or gas breakthrough.

c. Recovery Factors: Reservoir heterogeneity directly impacts the ultimate recovery of hydrocarbons. Variations in permeability and fluid saturations affect the efficiency of fluid displacement and the sweep of hydrocarbons within the reservoir. By simulating well tests in heterogeneous reservoirs, engineers can estimate recovery factors and assess the effectiveness of various production techniques, such as water flooding or enhanced oil recovery methods.

4. Optimization and Reservoir Management:

Numerical simulation models enable engineers to optimize production strategies and reservoir management in the presence of heterogeneity. By simulating different well test scenarios and considering reservoir heterogeneity, engineers can evaluate the impact of various factors, such as well placement, completion design, and injection rates, on reservoir performance.

In addition, numerical models make it easier to conduct sensitivity assessments, which help pinpoint the primary reservoir characteristics that have an impact on well test results. Engineers are able to evaluate the impact of these parameters on flow behavior, pressure responses, and recovery factors by systematically changing the values of these parameters. This knowledge helps drive decision-making, which enhances reservoir management and contributes to mitigating hazards associated with variability in reservoirs.

In conclusion, the heterogeneity of the reservoir is an important consideration in well testing as well as reservoir performance. Including reservoir heterogeneity, simulating well testing, and gaining thorough insights into flow behavior, pressure responses, and recovery variables are all things that may be accomplished with the help of powerful tools like numerical simulation models. Engineers are able to optimize production techniques, analyze reservoir connectivity, determine optimum flow routes, and make educated judgments on reservoir management and field development when they take into account heterogeneity.

# Fluid Displacement and Multiphase Flow

In the process of testing wells, fluid displacement and multiphase flow play an extremely important part, particularly when working with reservoirs that contain numerous fluids at once, such as oil, water, and gas. Numerical simulation models are strong tools that can accurately simulate and anticipate the behavior of these fluids during well testing by taking into consideration a variety of complicated phenomena. These models can simulate and predict the behavior of these fluids with high levels of precision. In the context of well testing simulation, the following is a comprehensive explanation of the most important aspects linked to fluid displacement and multiphase flow:

1. Multiphase Flow: Multiphase flow refers to the simultaneous flow of different fluids (e.g., oil, water, and gas) within the reservoir. This occurs because reservoirs are often composed of multiple layers or zones with different fluid saturations. Numerical simulation models are capable of handling multiphase flow by solving a set of equations known as the conservation equations, which describe the flow and transport of each fluid phase. These equations consider fluid properties, such as density, viscosity, and compressibility, along with the reservoir properties, such as permeability, porosity, and saturation.

2. Capillary Pressure: Capillary pressure is the pressure difference across the interface between two immiscible fluids in porous media. It plays a significant role in fluid displacement during well testing. Numerical simulation models incorporate capillary pressure curves derived from laboratory experiments or empirical correlations to capture the effects of capillary pressure on multiphase flow. These curves describe the relationship between the fluid saturations and the capillary pressure, influencing fluid distribution and movement within the reservoir.

3. Relative Permeability: Relative permeability is a fundamental concept in multiphase flow that describes the ability of each fluid phase to flow through the porous media. It represents the fractional flow of a particular fluid phase as a function of its saturation. Numerical simulation models utilize relative permeability curves, obtained through laboratory measurements or empirical correlations, to characterize the flow behavior of each fluid phase. These curves provide information on how the presence of one fluid phase affects the permeability and flow of other phases, influencing the displacement patterns during well testing.

4. Fluid Displacement Mechanisms: Fluid displacement mechanisms refer to the physical processes involved in displacing one fluid phase with another. They can be categorized into primary, secondary, and tertiary mechanisms. Numerical simulation models account for these mechanisms to accurately simulate the displacement of fluids during well testing. Some common displacement mechanisms include viscous fingering, capillary trapping, gravity segregation, and fluid front instabilities. By capturing these mechanisms, engineers can evaluate the efficiency of different production and injection strategies, understand the movement of fluids within the reservoir, and estimate the ultimate recovery of hydrocarbons.

5. Phase Behavior: The behavior of fluids within the reservoir is influenced by the phase behavior, which describes the interactions between different fluid phases. Numerical simulation models incorporate equations of state and phase behavior models to account for the changes in fluid properties under different pressure and temperature conditions. These models consider factors such as phase equilibrium, phase compositions, and phase densities. Accurate phase behavior modeling enables the simulation of

complex scenarios, such as fluid phase transitions (e.g., gas condensation) or compositional variations during well testing.

Engineers can obtain a full understanding of fluid displacement and multiphase flow during well testing by adding these characteristics into numerical simulation models. This enables the engineers to make more informed decisions. They are able to evaluate the effectiveness of various production and injection tactics, comprehend the flow and distribution of fluids inside the reservoir, and estimate the amount of hydrocarbons that may be recovered thanks to the models. It is essential to have this knowledge in order to make educated choices about the management of reservoirs, the optimization of production, and the planning of field development.

# Sensitivity Analysis and Optimization

While applying numerical simulation models for well testing, sensitivity analysis and optimization are two processes that are absolutely necessary. These techniques make it possible for engineers to comprehend how various parameters influence the test responses and to optimize the design of well testing in order to improve reservoir characterisation. In the context of well testing, here is a full explanation of sensitivity analysis and optimization:

1. Sensitivity Analysis:

Sensitivity analysis involves systematically varying input parameters to assess their impact on the output or response of the simulation model. In the case of well testing, engineers typically explore the sensitivity of various factors, including reservoir properties, well configurations, and fluid properties. The goal is to understand which parameters significantly influence the test responses and how they interact with each other. Some key considerations for sensitivity analysis are:

a. Reservoir Properties:

- Permeability: Varying permeability values across the reservoir can have a significant impact on pressure responses during well testing. Sensitivity analysis helps determine the sensitivity of pressure behavior to changes in permeability values, identify flow paths, and evaluate the connectivity of the reservoir.

- Porosity: Different porosity values affect fluid storage capacity and flow behavior. Sensitivity analysis helps assess how variations in porosity impact pressure responses and fluid displacement.

- Reservoir Thickness: Changes in reservoir thickness can influence pressure responses, flow rates, and ultimate recovery. Sensitivity analysis helps understand the sensitivity of these parameters and their influence on well test results.

b. Well Configurations:

 - Well Type and Placement: Sensitivity analysis allows engineers to assess the impact of well types (vertical, horizontal, multilateral) and their locations (depth, distance) on pressure responses and flow behavior. It helps optimize well placement to maximize production rates and recovery.

- Wellbore Storage: The storage capacity of the wellbore affects pressure behavior during well tests. Sensitivity analysis helps determine the sensitivity of pressure responses to wellbore storage and optimize its impact.

c. Fluid Properties:

- Fluid Mobility: Sensitivity analysis helps evaluate the impact of fluid properties (viscosity, compressibility) on well test responses. It helps understand the influence of fluid mobility on flow behavior, pressure profiles, and displacement efficiency.

In sensitivity analysis, one parameter is methodically changed while the others remain unchanged, and the changes in the well test responses that occur as a result are observed and analyzed. This process provides insights into which parameters have the most significant influence on the test results, allowing engineers to prioritize their efforts for accurate reservoir characterization. This process also provides insights into which parameters have the most significant influence on the test results.

2. Optimization:

Optimization involves the systematic selection of the most appropriate values for various parameters to achieve specific objectives. In the context of well testing, optimization aims to design the test duration, flow rates, and pressure measurement locations that will yield the most informative and accurate results. Key aspects of optimization include:

a. Test Duration: The duration of a well test influences the quality and quantity of data obtained. Optimizing the test duration involves considering factors such as the reservoir's response time, noise levels, and desired accuracy. Numerical simulation models can simulate various test durations, and engineers can evaluate the sensitivity of pressure responses to test duration to determine the optimal duration that provides sufficient data without excessive costs or time.

b. Flow Rates: The flow rates during well testing affect pressure responses and fluid displacement. Optimizing flow rates involves assessing the sensitivity of test results to different rates, considering factors such as reservoir heterogeneity and fluid behavior. Numerical simulation models enable engineers to simulate different flow rate scenarios and evaluate their impact on pressure responses, helping determine the optimal flow rates that provide the most valuable information about the reservoir.

c. Pressure Measurement Locations: The placement of pressure sensors during well testing plays a critical role in capturing the behavior of the reservoir. Optimization involves identifying the most appropriate locations for pressure measurements, considering factors such as reservoir heterogeneity, flow paths, and data acquisition capabilities. Numerical simulation models allow engineers to simulate pressure responses at different locations and evaluate their sensitivity to better understand the optimal placement of pressure sensors.

The goal of optimization is to maximize resource recovery while simultaneously reducing uncertainty and improving the accuracy of reservoir characterization.

Engineers are able to create well tests that provide the most valuable insights into the behavior of reservoirs by methodically evaluating a variety of characteristics and the influence those parameters have on the responses of the well tests.

To summarize, sensitivity analysis and optimization in numerical simulation models for well testing include making systematic changes to parameters in order to gain an understanding of how those changes affect the test responses. This method assists in determining the aspects that have the most impact and optimizes the design of well tests, which includes the test length, flow rates, and pressure measurement locations. Engineers are able to improve reservoir characterisation, minimize uncertainty, and make educated decisions for successful reservoir management and hydrocarbon recovery when they undertake sensitivity analyses and optimize.

## Field-Scale Behavior

The modeling of well tests that takes into account the entirety of the reservoir, including all of its geological characteristics, well configurations, and reservoir borders, is referred to as field-scale behavior. This method offers a more in-depth analysis of the behavior of the reservoir and enables engineers to investigate the interactions between several wells as well as the effects of a variety of parameters on production and injection rates, pressure responses, and ultimate recovery.

Instead of concentrating simply on individual wells, it is important to have a comprehensive understanding of the behavior of the reservoir as a whole while carrying out tests on the wells. The assumptions on reservoir behavior are frequently simplified by analytical models, for as by presuming that the reservoir is homogeneous or by ignoring the impact of reservoir boundaries. In spite of the fact that these simplifications can be helpful for quick estimations, they are unable to reflect the complexity that is present in reservoirs in the actual world.

On the other hand, numerical simulation models can represent the reservoir as a three-dimensional grid of cells, with each cell having a unique set of properties such as its permeability, porosity, and fluid saturations. This type of representation is known as a "cell-based" model. These models solve the governing equations that characterize fluid flow in porous media using numerical methods. Some examples of these equations include Darcy's law and the continuity equation.

Engineers are able to take into account the interactions between numerous

142

wells by modeling well testing at the field scale. This is particularly important in reservoirs that have complicated well configurations. For instance, when there are many producing wells, the pressure decline in one well can affect the pressure and fluid flow in nearby wells owing to pressure interference. This can occur when there are too many wells producing oil or gas. In a similar manner, the pressure that builds up in one well during injection well tests can have an effect on how the wells in the surrounding area behave.

In addition, the limits of the reservoir and the geological characteristics within it play a considerable effect in the behavior of the reservoir. Engineers are able to examine the impact that these elements have on well test responses by using numerical simulation models since these models can incorporate these parameters into the simulation. The existence of fault zones, stratigraphic differences, and other geological heterogeneities, for example, can have a substantial impact on fluid flow patterns, pressure distributions, and production or injection rates. Engineers are able to acquire insights into how the architecture of the reservoir affects the overall performance if they make these elements a part of the simulation and take them into account.

Moreover, numerical simulation methods are able to simulate the movement of fluid across the borders of reservoirs. The impermeable formations or aquifers that often constitute the margins of reservoirs can have a considerable influence on the flow of fluids that occur within the reservoir itself. Engineers are able to determine the influence that these external elements have on the responses of well tests by taking into account the behavior at the borders of the system. These factors include pressure communication with neighboring formations and water input from aquifers

Engineers are able to investigate a variety of scenarios and determine the impact of various elements on reservoir performance when they undertake simulations of well tests at the field scale. They are able to determine the effects of well spacing and completion procedures, examine the ideal placement of wells, and identify possible regions of enhanced production or injection. It is essential to have this knowledge in order to make educated choices about the management of reservoirs, the optimization of production, and the planning of field development.

It is essential to keep in mind that carrying out field-scale numerical simulations calls for the collection of specific reservoir characterisation data, which may include rock parameters, fluid properties, and well data. The dependability of these input parameters determines the quality of the simulation results, which in turn determines how accurate the simulation results are. In addition, the computational demands of field-scale simulations can be rather substantial, necessitating the use of high-performance computing resources in conjunction with effective numerical algorithms.

To summarize, field-scale behavior simulation using numerical models enables engineers to evaluate the impact of numerous parameters on well test results, consider the implications of reservoir boundaries and geological features, and analyze the interactions between several wells. Engineers can acquire a more thorough understanding of reservoir behavior by simulating the reservoir as a whole. This leads to improved reservoir management methods and greater production/injection optimization.

# Future Performance Prediction

The ability to forecast how a reservoir will function in the future is an essential part of reservoir engineering, and numerical simulation models are very important in this regard. Because of these models, engineers are able to simulate and make predictions regarding the behavior of a reservoir that extends beyond the time frame of a well test. Engineers are able to confirm the accuracy and dependability of the model by calibrating the simulation using historical well test data. Once the model has been validated, it can then be used to predict future reservoir performance. Using numerical simulation models to forecast future performance involves the following steps, as well as other important considerations, as outlined below:

1.  Model Calibration: To forecast future reservoir performance, engineers start by developing a numerical simulation model that represents the reservoir's geological and fluid properties. This involves creating a grid system that discretizes the reservoir into cells or blocks, each with assigned properties such as permeability, porosity, and initial fluid saturations. The model should also consider the presence of geological features, heterogeneities, and any existing production/injection wells.

2.  Historical Well Test Data: The next step is to gather historical well test data, which includes measurements of pressure, flow rates, and other relevant parameters during previous well tests. This data serves as a benchmark for calibrating the simulation model. By adjusting key reservoir and fluid properties within reasonable ranges, engineers aim to match the model's predicted behavior with the observed data from the well tests.

3.  History Matching: History matching is the process of adjusting the simulation model's parameters to minimize the mismatch between the predicted and observed well test data. Engineers iteratively

145

modify parameters such as reservoir permeability, porosity, and fluid properties to achieve a close match. This iterative process involves running multiple simulations and comparing the model's output with the historical data until an acceptable level of agreement is achieved.

4. Validation: Once the simulation model is calibrated and matches the historical well test data reasonably well, it is considered validated. The validation ensures that the model is representative of the reservoir's behavior and can be relied upon for future predictions. However, it's important to note that no model is perfect, and there will always be some level of uncertainty in the predictions.

5. Future Performance Forecasting: With a validated simulation model in hand, engineers can use it to forecast future reservoir performance. They can simulate various scenarios by adjusting production/injection rates, well configurations, or field development strategies. The model calculates pressure responses, flow rates, and fluid distributions over time based on the chosen scenarios.

6. Long-Term Productivity Estimation: By running the simulation model for an extended period, typically spanning several years or decades, engineers can estimate the long-term productivity of the reservoir. They can assess the production decline rates, water or gas breakthrough, and changes in fluid saturations over time. These insights help in understanding the reservoir's overall behavior and determining its ultimate recovery potential.

7. Optimal Field Development Strategies: Numerical simulation models also assist in identifying optimal field development strategies. Engineers can evaluate different well placement options,

production/injection rates, and reservoir management techniques using the simulation model. By comparing the performance of various scenarios, they can determine the most efficient and cost-effective approach to maximize hydrocarbon recovery from the reservoir.

8. Uncertainty Analysis: It's crucial to recognize that uncertainty exists in reservoir performance prediction. Engineers should conduct uncertainty analyses by considering different plausible scenarios and varying uncertain parameters within a range. This helps in quantifying the uncertainty associated with the predictions and provides a range of potential outcomes.

9. Iterative Process: Reservoir performance prediction using numerical simulation models is an iterative process that involves continuous monitoring and updating of the model as new data becomes available. As additional production data, pressure data, or well test data is acquired, engineers can recalibrate and refine the model to improve its accuracy and reliability.

Engineers are able to obtain significant insights into the behavior of a reservoir over the long term through the utilization of numerical simulation models for the purpose of future performance prediction. Because of this knowledge, they are able to make educated decisions regarding field development plans, reservoir management methods, and production optimization, which eventually leads to oil and gas operations that are more efficient and cost-effective.

# Field Applications

In the oil and gas sector, test analysis is extremely important for a number of different reasons, including assessing reservoir attributes, projecting production rates, optimizing well placement, and analyzing reservoir performance. Some examples of how test analysis can be applied in the following domains are as follows:

1. Estimating Initial Reservoir Pressure:
   - Pressure Build-Up Tests: By conducting pressure build-up tests, operators can measure the pressure response of the reservoir after shutting in a well. Analyzing the pressure data obtained from these tests helps estimate the initial reservoir pressure, which is vital for reservoir characterization and production forecasting.

2. Forecasting Production Rates:
   - Production Decline Analysis: By analyzing production decline curves, which represent the rate at which production decreases over time, engineers can estimate future production rates. This analysis helps in predicting reservoir performance, optimizing production strategies, and determining the economic viability of a field.

3. Optimizing Well Placement:
   - Well Testing and Well Log Analysis: By conducting well tests and analyzing the data obtained, operators can assess reservoir properties such as permeability, porosity, and fluid mobility. Combining this information with well log data (geophysical measurements), engineers can optimize well placement to target the most productive zones within the reservoir.

4.  Evaluating Reservoir Performance:
    - Pressure Transient Analysis: This analysis involves interpreting pressure data obtained from well tests, such as drawdown and buildup tests. By studying pressure transient behavior, engineers can evaluate reservoir performance, estimate reservoir parameters (e.g., permeability, skin factor), and identify potential issues like wellbore damage or reservoir boundaries.

5.  Reservoir Characterization:
    - Well Testing and Production Data Analysis: Integration of well test data with production data allows engineers to characterize the reservoir and estimate properties such as reservoir permeability, porosity, and fluid saturation. This information is vital for reservoir modeling, field development planning, and determining reservoir reserves.

6.  Enhanced Oil Recovery (EOR) Evaluation:
    - EOR Pilot Tests: Before implementing enhanced oil recovery techniques like water flooding, gas injection, or chemical flooding on a larger scale, operators often conduct pilot tests. Analyzing the results from these tests helps evaluate the effectiveness of the EOR method, assess reservoir response, and determine the feasibility of full-scale implementation.

These examples demonstrate the broad application of test analysis in the oil and gas industry, highlighting its importance in understanding reservoir behavior, optimizing production strategies, and making informed decisions to enhance recovery and maximize economic returns.

# Estimating Initial Reservoir Pressure

In the oil and gas industry, one of the most important tasks is to estimate the initial pressure in the reservoir. This is because it not only helps with reservoir characterization, production forecasting, and planning well interventions, but it also provides valuable information about the potential of the reservoir. In order to provide an accurate assessment of the initial pressure in the reservoir, pressure build-up tests are typically done.

During a pressure build-up test, a well is plugged off for an allotted amount of time so that the pressure in the reservoir can gradually increase to its full capacity. The pressure response over time is measured using pressure gauges that are situated both within the wellbore and in the surrounding observation wells. In order to conduct an analysis of these pressure readings, it is necessary to plot the pressure accumulation curve and then use mathematical models in order to estimate the original pressure in the reservoir.

The Horner plot is a well-known mathematical model that is utilized frequently for the analysis of pressure build-up. The plot known as the Horner plot is a semi-logarithmic representation of the prior

$$H(t) = (Pwf - Pi) / (t + m)$$

Where:

- $H(t)$ is the Horner plot function

- $Pwf$ is the wellbore flowing pressure (measured during the build-up test)

- $Pi$ is the initial reservoir pressure

- $t$ is the shut-in time

- m is a constant related to the reservoir properties and fluid flow behavior

A straight line can be obtained by graphing the Horner plot function H(t) against the shut-in time (t) on a semi-logarithmic scale. This will provide the straight line. The value of the constant m is represented by the slope of this line, and an estimate of the initial reservoir pressure can be derived from the point where the y-axis intersects the slope (Pi).

Now, let's go through an example to demonstrate what I mean by the estimation.

Example:

Suppose a pressure build-up test was conducted on a well, and the following data was obtained:

Shut-in time (t):

0 hours: 0 psi

1 hour: 400 psi

2 hours: 900 psi

4 hours: 1700 psi

8 hours: 3200 psi

Wellbore flowing pressure (Pwf): 3200 psi

151

We will use this data to estimate the initial reservoir pressure (Pi) using the Horner plot.

First, we calculate the Horner plot function H(t) for each shut-in time:

$H(0) = (3200 - Pi) / (0 + m)$

$H(1) = (3200 - Pi) / (1 + m)$

$H(2) = (3200 - Pi) / (2 + m)$

$H(4) = (3200 - Pi) / (4 + m)$

$H(8) = (3200 - Pi) / (8 + m)$

Next, we plot H(t) against t on a semi-logarithmic scale. The resulting plot should be a straight line:

```
```

H(t)

|

|      x

|        \

|         \

|          \

|           \

```
|          \
|_____
 0   1   2   4   8   t
```
```
```

From the plot, we can observe that the data points roughly form a straight line. We need to determine the slope of this line and the intercept on the y-axis.

Let's consider two data points, $(t1, H(t1)) = (2, H(2))$ and $(t2, H(t2)) = (8, H(8))$. The slope of the line is given by:

Slope = $(H(t2) - H(t1)) / (t2 - t1)$

Substituting the values from the plot, we get:

Slope = $[(3200 - Pi) / (8 + m) - (3200 - Pi) / (2 + m)] / (8 - 2)$

Now, we can calculate the intercept on the y-axis, which corresponds to the initial reservoir pressure (Pi). For this, we choose any data point on the line and substitute the values into the Horner plot equation. Let's choose $(t3, H(t3)) = (1, H(1))$:

$H(t3) = (3200 - Pi) / (1 + m)$

153

Solving for Pi, we get:

$$Pi = 3200 - H(t3) * (1 + m)$$

Finally, we can solve for Pi by simultaneously solving the equations for slope and Pi:

$$Slope = [(3200 - Pi) / (8 + m) - (3200 - Pi) / (2 + m)] / (8 - 2)$$

$$Pi = 3200 - H(t3) * (1 + m)$$

We are able to estimate the starting reservoir pressure (Pi) based on the pressure build-up test data if we repeatedly solve these equations using numerical approaches such as fitting using least squares.

In the real world, extra considerations and corrections may be necessary based on the properties of the reservoir, the behavior of the fluid, and other aspects; however, this example only presents a simplified summary of the situation. Please keep this in mind.

# Forecasting Production Rates

In the oil and gas business, production decline analysis is a technique that is utilized frequently in order to anticipate future production rates and evaluate the performance of reservoirs. It requires performing an analysis of past production data in order to comprehend the pattern of decline and anticipate the production profile for the future.

The optimization of production tactics, the planning of field development, and the determination of whether or not a field is economically viable all require this information.

Graphically representing the production rate as a function of time, the production decline curve shows how the rate of production has been decreasing through time. In most cases, it adheres to a pattern that is characterized by an initial production rate that is high and progressively decreases over the course of time. The decline may be brought on by a number of different variables, including fluid characteristics, reservoir heterogeneity, fluid depletion, and well performance.

Arps' decline curve equation is a model that is frequently utilized for the analysis of production decline. This model depicts the behavior of production decrease based on empirical evidence. This is the equation that needs to be solved:

$$Q(t) = Q(i) / [(1 + bD(t - t(i)))^{(1 / b)}]$$

Where:

- $Q(t)$ is the production rate at time t

- Q(i) is the initial production rate at time t(i)

- D is the decline rate constant

- t is the time at which the production rate is calculated

- t(i) is the initial time when production started

- b is the decline exponent

The decline exponent (b) determines the shape of the decline curve and reflects the dominant decline mechanism. It can take different values depending on the reservoir characteristics and production mechanisms. Common values for b are:

- b = 0: Exponential decline

- b = 1: Hyperbolic decline

- b = 2: Harmonic decline

To estimate the decline parameters (Q(i), D, b), engineers typically perform regression analysis on historical production data. By fitting the Arps' equation to the production data, they can determine the best-fit parameters that represent the decline behavior of the reservoir.

Let's consider a worked example to illustrate the production decline analysis:

Suppose we have historical production data for an oil well over a period of 5 years. The production rates at different time intervals are as follows:

Year 1: 1,000 barrels per day (bpd)

Year 2: 800 bpd

Year 3: 600 bpd

Year 4: 400 bpd

Year 5: 300 bpd

We want to estimate the future production rate at Year 6 using the Arps' decline curve analysis.

Step 1: Determine the initial production rate ($Q(i)$) and initial time ($t(i)$).

  - $Q(i) = 1,000$ bpd (from Year 1)

  - $t(i) = 0$ (assuming production started at the beginning of Year 1)

Step 2: Calculate the decline rate constant (D).

  - $D = (Q(i) - Q(t)) / (Q(t) * (t - t(i)))$

  - For simplicity, let's calculate D using the data from Year 5 and Year 4:

  $D = (300$ bpd $- 400$ bpd$) / (400$ bpd $* (5 - 4))$

  $D = -0.25$

Step 3: Determine the decline exponent (b).

  - To estimate the decline exponent, engineers typically perform regression analysis on historical data or use prior knowledge based on similar reservoirs. Let's assume $b = 1$ for this example.

157

Step 4: Calculate the future production rate at Year 6 ($Q(t)$).

 - $t = 6$ (Year 6)

 - $Q(t) = Q(i) / [(1 + bD(t - t(i)))^{(1 / b)}]$

 - $Q(t) = 1,000$ bpd $/ [(1 + 1 * -0.25 * (6 - 0))^{(}$

$1 / 1)]$

 - $Q(t) = 1,000$ bpd $/ (0.5^{1})$

 - $Q(t) = 2,000$ bpd

Based on the analysis, the estimated production rate at Year 6 is 2,000 barrels per day.

It is essential to understand that production decline analysis provides a forecast that is founded on past data as well as assumptions regarding the pattern of the decrease. The decline curve model does not take into account all of the unknowns and factors that could affect output rates, therefore actual production rates could end up being different. In order to improve production forecasts and make production strategies as effective as possible, continuous monitoring of decline parameters and frequent reevaluation of those parameters are required.

# Enhanced Oil Recovery (EOR) Evaluation

Enhanced Oil Recovery, often known as EOR, refers to the processes that are used to increase the amount of oil that is extracted from reservoirs above and beyond what is possible utilizing primary and secondary recovery techniques. EOR approaches are often evaluated for their efficacy, the response of the reservoir, and the feasibility of full-scale application by operators before the techniques are put into practice on a wider scale. Throughout the course of these pilot tests, fluids are injected into the reservoir, and the effect those fluids have on the production of oil is monitored.

Water flooding is a typical form of enhanced oil recovery (EOR) that includes injecting water into the reservoir in order to move the oil towards the producing wells. By assessing the outcomes of the pilot test, specifically by evaluating the waterflood performance and estimating the incremental oil recovery, it is possible to determine how effective water flooding is.

In order to assess how well a waterflood pilot test performed, it is necessary to take into account various critical characteristics, including the following:

1.  Injection Rate (Qinj): The rate at which water is injected into the reservoir, typically measured in barrels per day (bpd).

2.  Production Rate (Qprod): The rate at which oil is produced from the wells affected by the pilot test, also measured in bpd.

3.  Cumulative Water Injected (CWPI): The total volume of water injected into the reservoir throughout the pilot test, usually measured in barrels (bbl).

4. Cumulative Oil Produced (COIP): The total volume of oil produced from the wells affected by the pilot test, measured in bbl.

5. Reservoir Pressure (P): The pressure within the reservoir, often measured in pounds per square inch (psi) or bars.

To assess the performance of the waterflood pilot test, the following parameters are commonly analyzed:

1. Injectivity Index (II): The injectivity index represents the ability of the reservoir to accept injected fluids and is calculated as the ratio of the injection rate to the pressure drop across the reservoir:

$$II = Qinj / \Delta P$$

Where $\Delta P$ is the pressure drop across the reservoir.

2. Waterflood Sweep Efficiency (SW): The sweep efficiency indicates the effectiveness of the injected water in displacing the oil within the reservoir. It is calculated as the ratio of the cumulative oil produced to the oil that could be produced by displacing the original oil in place (OOIP):

$$SW = COIP / OOIP$$

The OOIP can be estimated based on the reservoir rock and fluid properties, reservoir volume, and formation evaluation data.

3. Waterflood Recovery Efficiency (RE): The recovery efficiency represents the fraction of the original oil in place that is ultimately recovered due to the waterflood process. It is calculated as the ratio of the cumulative oil produced to the OOIP:

RE = COIP / OOIP

Recovery efficiency can also be calculated based on the difference between the cumulative water injected and the cumulative oil produced:

RE = (CWPI - COIP) / OOIP

Alternatively, the recovery efficiency can be calculated based on the difference between the initial oil in place (IOIP) and the remaining oil in place (ROIP):

RE = (IOIP - ROIP) / IOIP

The IOIP and ROIP can be estimated using reservoir simulation or other analytical methods.

The performance of the waterflood pilot test can be evaluated by the operators through the examination of the injectivity index, the efficiency of the waterflood sweep, and the recovery efficiency. These characteristics give information on the reservoir's response to the waterflood, the capacity of

161

the injected water to displace oil, and the potential for enhanced oil recovery on a wider scale.

Let's consider a worked example to illustrate the evaluation of a waterflood pilot test:

Suppose a waterflood pilot test was conducted in a reservoir with the following data:

- Injection Rate (Qinj): 2,000 bpd

- Production Rate (Qprod):

 1,500 bpd

- Cumulative Water Injected (CWPI): 100,000 bbl

- Cumulative Oil Produced (COIP): 50,000 bbl

- Original Oil in Place (OOIP): 500,000 bbl

We can calculate the injectivity index, waterflood sweep efficiency, and recovery efficiency as follows:

1.  Injectivity Index (II):

Assuming a pressure drop ($\Delta P$) of 500 psi:

II = 2,000 bpd / 500 psi = 4 bpd/psi

2. Waterflood Sweep Efficiency (SW):

SW = 50,000 bbl / 500,000 bbl = 0.1 or 10%

3. Waterflood Recovery Efficiency (RE):

Using the cumulative oil produced:

RE = 50,000 bbl / 500,000 bbl = 0.1 or 10%

Using the difference between cumulative water injected and cumulative oil produced:

RE = (100,000 bbl - 50,000 bbl) / 500,000 bbl = 0.1 or 10%

Using the difference between initial oil in place and remaining oil in place:

Assuming the remaining oil in place (ROIP) is 300,000 bbl:

RE = (500,000 bbl - 300,000 bbl) / 500,000 bbl = 0.4 or 40%

163

We are able to make sense of the findings of the waterflood pilot test now that we have the computed data. An injectivity index of 4 bpd/psi demonstrates that the reservoir has a reasonable ability to receive injected fluids since it shows that the reservoir can accept the injected fluids. The fact that just ten percent of the oil in the reservoir was removed by the waterflood suggests that only a little portion of the oil was actually displaced by the water. The waterflood recovery efficiency of 10% or 40% (depending on the technology utilized) implies that there is the possibility for further oil recovery through the waterflood process.

These values can be used as a basis for decision-making on the practicability and effectiveness of implementing a waterflood on a large scale in the reservoir. In order to evaluate the long-term performance of the waterflood project and determine whether or not it is economically viable, additional research would be necessary. This research would include reservoir modeling studies, economic evaluations, and sensitivity analyses.

Note that the example that has been provided is merely for illustrative purposes, and that the actual evaluation of a waterflood pilot test would involve a more in-depth data analysis that takes into account a variety of factors, including reservoir heterogeneity, fluid properties, and geologic characteristics.

# Future Trends

In recent years, there have been considerable developments in the field of well test analysis. These advancements have been driven by the integration of sophisticated data analytics, methods of machine learning, and real-time well testing technologies. These new developments have the potential to improve the effectiveness and precision of reservoir characterization, thereby delivering invaluable insights into the behavior and characteristics of hydrocarbon reservoirs. Let's get into more depth about each of these emerging tendencies:

1. Advanced Data Analytics: Because of the rise of big data and the availability of huge volumes of data from well-designed experiments, the utilization of sophisticated data analytics methods has become absolutely necessary in order to extract relevant information and trends. When applied to well test data, data analytics methods including statistical analysis, pattern recognition, and data mining can be used to detect trends, anomalies, and correlations in the data. These investigations contribute to a better understanding of the dynamics of the reservoir, as well as to the estimation of the parameters of the reservoir and the optimization of production techniques.

2. Machine Learning: The application of machine learning techniques to the analysis of well tests has seen substantial progress. Machine learning approaches can find complicated patterns and correlations that may not be obvious using standard analysis methods. These patterns and relationships can be found by training models on vast datasets containing historical well test data. After that, you may use these models to make predictions about the features of the reservoir, optimize your well testing procedures, and even spot early warning signals of damage or abnormalities in the reservoir. Real-time analysis of well data can also benefit from the application

of machine learning because it enables automated interpretation and decision-making skills.

3. Real-time Well Testing Techniques: The traditional procedures for testing wells entailed gathering data over extended periods of time and then assessing it after the test had been completed. Real-time well testing procedures, on the other hand, make it possible to conduct ongoing monitoring and analysis of a well's behavior while the testing is being done. This makes it possible to immediately identify transient effects, locate reservoir boundaries early on, and make rapid adjustments to testing conditions. Real-time monitoring systems, when coupled with automated analytical algorithms, offer timely insights, make it easier to make decisions based on data, and eventually improve reservoir characterisation and optimize production methods.

4. Integration of Measurements: The integration of multiple measurements received from well testing is beneficial to the process of characterisation of reservoirs. Combining, for example, the pressure, temperature, flow rate, and production data from multiple wells or sensors can provide a more in-depth understanding of the behavior and attributes of the reservoir. When these observations are integrated with more advanced analytical approaches, it is possible to identify complex reservoir features like as fractures, heterogeneities, and compartmentalization. These features are essential for correct reservoir characterization and modeling.

5. Uncertainty Quantification: The analysis of well tests must always include the step of quantifying uncertainty. The integration of uncertainty quantification approaches into reservoir characterization workflows is becoming an increasingly popular emerging trend. The analysis of the well test can be made more robust and reliable by taking into account the uncertainties

associated with the input parameters, the quality of the data, and the modeling assumptions. In order to evaluate and propagate uncertainty, modeling methods like Monte Carlo simulation, Bayesian inference, and ensemble modeling are utilized. As a result, a more accurate depiction of how the reservoir behaves is produced, and decision-making procedures are made more effective.

In a nutshell, the incorporation of cutting-edge data analytics, machine learning, and real-time well testing procedures into well test analysis has the potential to revolutionize reservoir characterization. These new trends allow for more efficient and accurate assessment of reservoir attributes, enhance the understanding of complicated reservoir behavior, optimize production techniques, and eventually lead to improved hydrocarbon recovery and field development.

# Appendices

These formulae and laws are foundational in reservoir engineering and are used for various calculations and analyses related to reservoir characterization, production forecasting, well performance evaluation, and reservoir management.

## Formulae

Darcy's Law: $Q = -kA\Delta P/\mu L$, where Q is the flow rate, k is the permeability, A is the cross-sectional area, $\Delta P$ is the pressure drop, $\mu$ is the fluid viscosity, and L is the length of the flow path.

Material Balance Equation: $PVI = (N\varphi/5.615)(Boi - Bgi) + (N\varphi/5.615)(Bt - Bt^*) + (N\varphi/5.615)(Bp - Bp^*)$, where PVI is the cumulative production divided by the initial hydrocarbon pore volume, $N\varphi$ is the net formation volume factor, Boi is the initial oil formation volume factor, Bgi is the initial gas formation volume factor, Bt is the total compressibility, Bt* is the change in total compressibility, Bp is the formation volume factor for produced oil, and Bp* is the change in formation volume factor for produced oil.

Buckley-Leverett Equation: $\partial(\varphi Sg)/\partial t + \partial(Qo)/\partial x = 0$, where $\varphi$ is the porosity, Sg is the gas saturation, Qo is the oil flow rate, t is time, and x is the distance.

Archie's Law: $Rt = a * Rwa * \varphi^m$, where Rt is the formation resistivity factor, a is a constant, Rwa is the water saturation resistivity, $\varphi$ is the porosity, and m is the cementation exponent.

Material Balance for Gas Reservoirs: GFGI = (Bg - Bgi) + (Np/Nfo)(Bo - Boi) + (Nw/Nfw)(Bw - Bwi), where GFGI is the cumulative produced gas divided by the initial gas in place, Bg is the gas formation volume factor, Np is the cumulative produced oil, Nfo is the original oil in place, Bo is the oil formation volume factor, Boi is the initial oil formation volume factor, Nw is the cumulative produced water, Nfw is the initial water in place, Bw is the water formation volume factor, and Bwi is the initial water formation volume factor.

Well Productivity Index (PI): PI = (khμ)/(Bμ(Pwf - Pwfres)), where kh is the horizontal permeability, μ is the fluid viscosity, B is the formation volume factor, Pwf is the wellbore flowing pressure, and Pwfres is the reservoir pressure.

Reservoir Drive Mechanisms: Primary drive (natural reservoir energy), Secondary drive (water or gas injection), Tertiary drive (enhanced oil recovery techniques).

Decline Curve Analysis: Q(t) = Qi / (1 + Dt)^b, where Q(t) is the production rate at time t, Qi is the initial production rate, D is the decline rate, and b is the decline exponent.

Bubble Point Pressure Calculation: Pbp = Rsi / (Bg - Bgi) + Pn, where Pbp is the bubble point pressure, Rsi is the solution gas-oil ratio, Bg is the gas formation volume factor, Bgi is the initial gas formation volume factor, and Pn is the pressure at the depth of interest.

Oil Volumetric Calculation: V = Np * Bo, where V is the cumulative oil volume, Np is the cumulative oil production, and Bo is the oil formation volume factor.

Gas Volumetric Calculation: $V = Ng * Bg$, where V is the cumulative gas volume, Ng is the cumulative gas production, and Bg is the gas formation volume factor.

Water Volumetric Calculation: $V = Nw * Bw$, where V is the cumulative water volume, Nw is the cumulative water production, and Bw is the water formation volume factor.

Fluid Saturation Calculation: $S = (V / (V + Vp)) * 100$, where S is the fluid saturation, V is the volume of the fluid of interest, and Vp is the volume of the porous medium.

Relative Permeability Calculation: $Kr = (K - Krl) / (Krg - Krl)$, where Kr is the relative permeability, K is the effective permeability, Krl is the residual liquid permeability, and Krg is the residual gas permeability.

Pressure Transient Analysis: Well testing technique to analyze pressure behavior and infer reservoir properties using pressure data.

Wellbore Storage Analysis: Estimation of reservoir properties by analyzing the pressure response due to wellbore storage effects during well shut-ins.

Isothermal Compressibility Calculation: $Ct = 1 / B * (\partial B / \partial P)$, where Ct is the isothermal compressibility, B is the formation volume factor, and

$\partial B / \partial P$ is the derivative of formation volume factor with respect to pressure.

Skin Factor Calculation: $S = (Pwf - Pe) / (2 * \pi * kh * h * \mu)$, where S is the skin factor, Pwf is the wellbore flowing pressure, Pe is the external reservoir pressure, kh is the horizontal permeability, h is the net pay thickness, and $\mu$ is the fluid viscosity.

Reservoir Pressure Calculation: $P = Pe + (\Delta h * \varrho * g)$, where P is the reservoir pressure, Pe is the external pressure, $\Delta h$ is the depth difference, $\varrho$ is the fluid density, and g is the acceleration due to gravity.

Reservoir Fluid Density Calculation: $\varrho = \varrho w * Sw + \varrho o * (1 - Sw)$, where $\varrho$ is the fluid density, $\varrho w$ is the water density, $\varrho o$ is the oil density, and Sw is the water saturation.

Pressure Gradient Calculation: $\partial P / \partial h = \varrho * g$, where $\partial P / \partial h$ is the pressure gradient, $\varrho$ is the fluid density, and g is the acceleration due to gravity.

Oil Recovery Factor Calculation: $RF = (Np / Nfo) * 100$, where RF is the oil recovery factor, Np is the cumulative oil production, and Nfo is the original oil in place.

Gas-Oil Ratio Calculation: $R = Ng / Np$, where R is the gas-oil ratio, Ng is the cumulative gas production, and Np is the cumulative oil production.

171

Water-Oil Ratio Calculation: $R = N_w / N_p$, where $R$ is the water-oil ratio, $N_w$ is the cumulative water production, and $N_p$ is the cumulative oil production.

Material Balance for Water Drive Reservoirs: $WOR = (B_w - B_{wi}) / (B_o - B_{oi})$, where $WOR$ is the water-oil ratio, $B_w$ is the water formation volume factor, $B_{wi}$ is the initial water formation volume factor, $B_o$ is the oil formation volume factor, and $B_{oi}$ is the initial oil formation volume factor.

Pore Volume Calculation: $V_p = \varphi * A * h$, where $V_p$ is the pore volume, $\varphi$ is the porosity, $A$ is the cross-sectional area, and $h$ is the net pay thickness.

Injectivity Index Calculation: $II = (Q / (\Delta P * \mu))$, where $II$ is the injectivity index, $Q$ is the injection rate, $\Delta P$ is the pressure drop, and $\mu$ is the fluid viscosity.

Water Saturation Calculation: $S_w = (V_w / (V_w + V_o)) * 100$, where $S_w$ is the water saturation, $V_w$ is the cumulative water volume, and $V_o$ is the cumulative oil volume.

Minimum Miscibility Pressure Calculation: $MMP = (Z_g * P_c)$, where $MMP$ is the minimum miscibility pressure, $Z_g$ is the gas compressibility factor, and $P_c$ is the critical pressure.

Wellbore Flowing Pressure Calculation: $P_{wf} = P_e - (2 * \pi * kh * h * \mu * Q) / (\ln(r_w / r_e) + S)$, where $P_{wf}$ is the wellbore flowing pressure, $P_e$ is the external reservoir pressure, $kh$ is the horizontal permeability, $h$ is the net pay thickness, $\mu$ is the fluid viscosity, $Q$ is the flow rate, $r_w$ is the wellbore radius, re is the drainage radius, and S is the skin factor.

# Bibliography

Ahmed, T. (2010). Reservoir Engineering Handbook (4th ed.). Gulf Professional Publishing.

Craft, B. C., & Hawkins, M. F. (1991). Applied Petroleum Reservoir Engineering (2nd ed.). Prentice Hall.

Chen, Z. (2017). Reservoir Engineering: The Fundamentals, Simulation, and Management of Conventional and Unconventional Recoveries. Gulf Professional Publishing.

Raghavan, R. (2007). Principles of Enhanced Oil Recovery. Elsevier.

Lyons, W. C., & Plisga, G. J. (2010). Standard Handbook of Petroleum and Natural Gas Engineering (3rd ed.). Gulf Professional Publishing.

Mattar, L., & Manrique, E. (2017). Advanced Reservoir Engineering. Wiley.

Lake, L. W. (1989). Enhanced Oil Recovery. Prentice Hall.

Tehrani, D. H., & Proett, M. N. (2011). Modern Reservoir Engineering: A

Simulation Approach. Wiley.

Horne, R. N. (2018). Modern Well Test Analysis: A Computer-Aided Approach. Gulf Professional Publishing.

Odeh, A. S., & Poettmann, F. H. (1984). Fundamentals of Reservoir Engineering. Developments in Petroleum Science, 18.

Dake, L. P. (2001). Fundamentals of Reservoir Engineering (Vol. 8). Elsevier.

Lake, L. W., Johns, R. T., & Rossen, W. R. (2015). Enhanced Oil Recovery: Field Planning and Development Strategies. Gulf Professional Publishing.

Farouq-Ali, S. M., & Lake, L. W. (2003). Improved Oil Recovery by Surfactant and Polymer Flooding. Gulf Professional Publishing.

Lohrenz, J., & Watkins, R. W. (1992). Applied Reservoir Engineering. PennWell Books.

Ma, C., & Holditch, S. A. (2003). Reservoir Engineering: Guidelines for Practice. Society of Petroleum Engineers.

Lake, L. W., Johns, R. T., & Nur, A. (2019). The Practice of Reservoir Engineering. Elsevier.

Sami, M., & Islam, M. R. (2019). Reservoir Engineering: Principles and Applications. Wiley.

Pinczewski, W. V., & Miller, R. (2001). Integrated Reservoir Asset Management: Principles and Best Practices. PennWell Books.

Civan, F. (1996). Reservoir Formation Damage: Fundamentals, Modeling, Assessment, and Mitigation. Gulf Publishing Company.

Fanchi, J. R. (2001). Principles of Applied Reservoir Simulation (2nd ed.). Gulf Professional Publishing.

Tehrani, D. H., & Rezaee, M. R. (2014). Reservoir Engineering Handbook (4th ed.). Gulf Professional Publishing.

Orr Jr, F. M. (2007). Coalbed Methane: Principles and Practices. Gulf Professional Publishing.

Al-Lawati, M. (2011). Reservoir Engineering: Principles and Applications. LAP Lambert Academic Publishing.

Buchanan, E. T. (1991). Applied Petroleum Reservoir Engineering (3rd ed.). Prentice Hall.

Alomair, O. A. (2014). Reservoir Engineering: Practical Techniques and Tools. CreateSpace Independent Publishing Platform.

Lake, L. W., Chen, W. L., & Yortsos, Y. C. (2001). Carbonate Reservoir Characterization: A Geologic-Engineering Analysis (Part I). Elsevier.

Paluszny, A., & Zimmerman, R. W. (2011). Hybrid Finite-Discrete Element Modeling of Reservoir Behavior. Elsevier.

Ertekin, T., Abou-Kassem, J. H., & King, G. R. (2001). Basic Applied Reservoir Simulation. Society of Petroleum Engineers.

Tiab, D., & Donaldson, E. C. (2012). Petrophysics: Theory and Practice of Measuring Reservoir Rock and Fluid Transport Properties (3rd ed.). Gulf Professional Publishing.

Odeh, A. S. (1988). Petroleum Reservoir Engineering Practice. Marcel Dekker.

Islam, M. R., & Choudhury, I. A. (2009). Reservoir Engineering Handbook (2nd ed.). LAP Lambert Academic Publishing.

Valkó, P. P. (2008). Production Optimization Using Nodal Analysis (2nd ed.). Schlumberger.

Weber, R. R. (2001). Reservoir Simulation: Mathematical Techniques in Oil Recovery (2nd ed.). Society of Petroleum Engineers.

Li, C. (2005). Petroleum Production Engineering: A Computer-Assisted Approach. Elsevier.

Srivastava, R. K., & Civan, F. (2019). Reservoir Formation Damage: Fundamentals, Modeling, Assessment, and Mitigation (3rd ed.). Gulf Professional Publishing.

Tehrani, D. H. (2005). Modern Reservoir Flow and Well Transient Analysis. Elsevier.

Aziz, K., & Settari, A. (1979). Petroleum Reservoir Simulation (1st ed.). Elsevier.

Mattar, L., & Ramazanova, A. (2016). Reservoir Engineering Handbook (5th ed.). LAP Lambert Academic Publishing.

Chierici, G. L. (2013). Reservoir Simulation. Lap Lambert Academic Publishing.

Medley, G. H., & Lewis, D. (1996). Practical Reservoir Simulation. Gulf Professional Publishing.

Al-Shaalan, S. (2006). Modern Petroleum Reservoir Simulation: A Computer-Based Approach. Elsevier.

Odeh, A. S. (1991). Reservoir Engineering Handbook. Butterworth-Heinemann.

Reynolds, A. C. (1997). Applied Reservoir Engineering. Butterworth-Heinemann.

Ahmed, I., & McKinney, P. (2012). Enhanced Oil Recovery Field Case Studies. Gulf Professional Publishing.

Katz, D. L., & Lee, R. L. (1990). Natural Gas Engineering: Production and Storage. McGraw-Hill.

Ahmed, T., & McKinney, P. (2010). Advanced Reservoir Engineering. Gulf Professional Publishing.

Lake, L. W., & Walsh, M. P. (1996). A Practical Approach to Reservoir Simulation. PennWell Books.

Skjaeveland, S. M. (2004). Reservoir Simulation: Mathematical Techniques in Oil Recovery (2nd ed.). Society of Petroleum Engineers.

Veeken, P. C. (2007). The Use of Electrochemical Scanning Tunneling

Microscopy (ECTSM) in Petroleum Reservoir Engineering: Applied to the Study of Oil Recovery by Wettability Alteration. CRC Press.

Reynolds, A. C., & Ali, S. (2010). Reservoir Engineering Aspects of Waterflooding. Society of Petroleum Engineers.

Made in the USA
Coppell, TX
26 September 2023

22057810R00101